Mathematics 5

Practice in the Basic Skills

CW00841534

Contents

Notation

A Write the number shown on each abacus in: **a** figures **b** words.

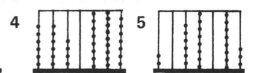

B Write in figures.

4 million	7 million	19 million	2 million	15 million
$\frac{1}{2}$ million	$\frac{1}{4}$ million	$\frac{3}{4}$ million	$\frac{1}{5}$ million	$\frac{1}{10}$ million
$3\frac{1}{4}$ million	$2\frac{3}{4}$ million	$6\frac{1}{2}$ million	$2\frac{3}{5}$ million	$4\frac{7}{10}$ million
0·1 million	0·4 million	0·7 million	0·2 million	0·6 million
3·3 million	2·5 million	1·8 million	3·4 million	6·9 million

C

		a		b		c	
Add 10 to:		a	402 341	b	160 592	c	999 999
Add 100 to:		a	634 801	b	502 916	c	1 271 214
Add 10 000 to:		a	437 216	b	50 439	c	207 986
Add 100 000 to:		a	921 741	b	843 216	c	1 941 276
Add 1000 to:		a	1 499 211	b	1 376 416	c	1 999 271

D Write the value in words of the underlined figure in each of these numbers.

43<u>5</u> 261	<u>2</u> 431 276	90<u>9</u> 214	<u>5</u>4 251
2<u>9</u>7 616	<u>1</u> 627 401	808 5<u>3</u>1	<u>7</u>56 394

E Write the number which is: **a** 1 less than **b** 10 less than **c** 100 less than each of these numbers.

$\frac{1}{2}$ million	$\frac{1}{4}$ million	$\frac{3}{4}$ million	300 000	350 000
4·2 million	2·1 million	8·3 million	470 000	610 000

Approximations

A Approximate these numbers to the nearest: **a** 10   **b** 100   **c** 1000.

| 54 729 | 13 269 | 202 435 | 60 987 | 120 696 |
| 444 909 | 38 451 | 899 823 | 54 163 | 914 429 |

B Approximate to the nearest: **a** hundredth   **b** tenth   **c** whole.

| 14·352 | 16·073 | 7·451 | 242·715 | 6·439 |
| 11·992 | 13·475 | 6·549 | 8·727 | 18·296 |

C Approximate to the nearest ten thousand.

| 429 000 | 516 214 | 1 421 216 | 801 216 | 34 921 |
| 1 427 500 | 350 001 | 2 069 350 | 609 215 | 414 916 |

D Approximate to the nearest million.

| 15 215 007 | 26 841 215 | 5 296 416 | 12 851 216 |
| 7 407 196 | 32 945 125 | 10 816 000 | 5 613 000 |

E Approximate to the nearest hundred thousand.

| 479 416 | 821 394 | 464 514 | 1 396 214 | 375 416 |
| 5 291 216 | 794 271 | 354 741 | 663 415 | 704 596 |

F Approximate to the nearest: **a** thousand   **b** ten thousand
c hundred thousand   **d** million

5 416 299   14 767 345   4 636 921   12 392 906   7 439 899   12 930 914

Addition and subtraction

A Find:

$156 + 40\,178 + \frac{1}{2}$ million

$20\,476 + 296\,751 + 500\,279$

$6006 + 23\,745 + 927\,651$

$1\cdot4$ million $+ 35\,427 + 253\,651$

$35 + 471 + 969$

$2\frac{3}{4}$ million $+ 500\,000 + 62\,753$

$\frac{1}{4}$ million $+ 947 + 4299$

$10\,800 + 39 + 976$

$6\cdot7$ million $+ \frac{1}{2}$ million $+ 300\,000$

$35\,000 + 98\,000 + 202\,998$

B

$\frac{3}{4}$ million $- 476\,296$

$217\,491 - 147\,281$

$3\cdot6$ million $- 2\,417\,399$

$500\,000 - \frac{1}{4}$ million

$247\,399 - 99\,541$

$3\,541\,000 - 2\,796\,395$

$1\,000\,005 - 868\,541$

$750\,000 - 0\cdot6$ million

$99\,000 - 87\,679$

$4\frac{3}{4}$ million $- 2\cdot7$ million

C

1 Find the total of two hundred and fifty thousand, $\frac{1}{2}$ million and seventeen thousand and sixty-six.

2 Find the difference between 431 621 and 808 471.

3 How much is $\frac{3}{4}$ million less than 943 541?

4 Find the sum of 42 714, 202 671, 5499 and 87.

5 How much greater than 492 741 is $\frac{1}{2}$ million?

6 43 216 plus 104 439 plus 99 549 plus 1 276 399.

7 1 439 992 subtract 953 899.

8 Twenty-four thousand minus sixteen thousand and nine.

9 Increase 4·7 million by 91 476.

10 Decrease 515 402 by 279 509.

11 1 435 629 plus 801 776 minus 927 598.

12 Take forty-one thousand nine hundred and fifty from the sum of three-quarters of a million and two hundred and fifty thousand and ninety-six.

Multiplication and division

A **1**

435	2459	3216	926	6027
×8	×6	×11	×5	×7

2 4359×9 $\qquad$ 9091×4 $\qquad$ $29\,361 \times 3$

B 375×10 $\qquad$ 426×10 $\qquad$ 1743×10 $\qquad$ $20\,549 \times 10$

5271×30 $\qquad$ 694×70 $\qquad$ 4358×90 $\qquad$ $16\,096 \times 80$

C

56	79	68	94	59	87
×16	×23	×54	×76	×17	×21

D

476	309	909	835	479	776
×32	×54	×27	×42	×67	×19

E 1075×13 $\qquad$ 1975×24 $\qquad$ 2461×92 $\qquad$ 8798×77

F $6\overline{)436}$ $\qquad$ $3\overline{)529}$ $\qquad$ $9\overline{)706}$ $\qquad$ $7\overline{)547}$ $\qquad$ $5\overline{)631}$ $\qquad$ $8\overline{)928}$

G $11\overline{)1271}$ $\qquad$ $7\overline{)2465}$ $\qquad$ $12\overline{)8325}$ $\qquad$ $9\overline{)7072}$ $\qquad$ $2\overline{)3546}$

H $12\overline{)20\,171}$ $\qquad$ $6\overline{)53\,542}$ $\qquad$ $4\overline{)67\,541}$ $\qquad$ $8\overline{)25\,372}$

I $16\overline{)96}$ $\qquad$ $31\overline{)87}$ $\qquad$ $15\overline{)59}$ $\qquad$ $23\overline{)93}$ $\qquad$ $13\overline{)47}$ $\qquad$ $21\overline{)63}$ $\qquad$ $27\overline{)82}$

J $19\overline{)132}$ $\qquad$ $29\overline{)257}$ $\qquad$ $21\overline{)189}$ $\qquad$ $32\overline{)305}$ $\qquad$ $28\overline{)267}$ $\qquad$ $36\overline{)354}$

K $17\overline{)294}$ $\qquad$ $33\overline{)825}$ $\qquad$ $16\overline{)751}$ $\qquad$ $22\overline{)683}$ $\qquad$ $35\overline{)937}$ $\qquad$ $27\overline{)541}$

L $18\overline{)1362}$ $\qquad$ $22\overline{)1541}$ $\qquad$ $51\overline{)1269}$ $\qquad$ $34\overline{)2515}$ $\qquad$ $19\overline{)1432}$

M $21\overline{)4240}$ $\qquad$ $17\overline{)3027}$ $\qquad$ $23\overline{)5632}$ $\qquad$ $18\overline{)7471}$ $\qquad$ $31\overline{)8415}$

N $4321 \div 26$ $\qquad$ $1090 \div 25$ $\qquad$ $3242 \div 35$ $\qquad$ $6070 \div 15$

Long multiplication

A **1** Multiply each number by: **a** 10   **b** 30   **c** 50   **d** 80.

| 19 | 26 | 25 | 73 | 51 | 46 | 83 | 92 | 54 | 31 |

2 Multiply each number by: **a** 100   **b** 400   **c** 700.

| 35 | 42 | 75 | 87 | 96 | 58 | 29 | 18 | 67 | 72 |

B **1**

175	269	727	586	392	826
×100	×100	×100	×100	×100	×100

2

341	505	679	921	541	432
×700	×200	×500	×300	×400	×600

C

247	903	362	456	535	929
×150	×170	×180	×190	×160	×130

D

563	395	992	156	716	832
×210	×610	×810	×710	×510	×910

E

222	405	817	283	635	883
×520	×470	×630	×860	×950	×770

F

455	631	754	429	987	654
×301	×406	×603	×502	×804	×205

G

928	333	679	831	524	246
×419	×724	×632	×868	×953	×383

116	472	669	785	357	598
×628	×717	×434	×569	×342	×853

Long division

A

| 15 | 27 416 | 14 | 34 327 | 17 | 53 416 | 16 | 24 137 |

| 17 | 41 206 | 18 | 39 502 | 13 | 60 271 | 19 | 59 328 |

B

| 26 | 40 321 | 22 | 57 621 | 25 | 35 471 | 23 | 27 909 |

| 28 | 61 342 | 27 | 92 147 | 24 | 51 151 | 21 | 42 106 |

C

| 36 | 26 329 | 33 | 61 112 | 35 | 25 324 | 32 | 91 351 |

| 37 | 42 151 | 39 | 52 001 | 34 | 80 191 | 38 | 17 629 |

D

| 21 | 43 152 | 19 | 14 351 | 37 | 45 161 | 38 | 39 124 |

| 29 | 77 321 | 25 | 19 720 | 26 | 38 262 | 38 | 77 066 |

E

| 141 | 967 | 115 | 627 | 105 | 629 | 123 | 767 |

F

| 154 | 1025 | 131 | 1246 | 183 | 1064 | 227 | 1946 |

G

| 161 | 1815 | 214 | 3416 | 181 | 5621 | 207 | 5152 |

H

| 115 | 14 767 | 121 | 75 403 | 211 | 67 693 |

| 165 | 42 600 | 205 | 64 251 | 231 | 82 196 |

I

$4372 \div 16$ $5419 \div 21$ $20\,013 \div 33$

$42\,161 \div 35$ $20\,421 \div 133$ $6261 \div 26$

$4216 \div 111$ $9091 \div 38$ $14\,251 \div 23$

J Divide:

4161 by 18   7241 by 36 $18\,647 \div 116$

12 031 by 65   24 716 by 29 $20\,307 \div 47$

Decimals

A Write the following vulgar fractions as decimals.

1 $\frac{12}{100}$ $\frac{3}{10}$ $\frac{17}{1000}$ $\frac{3}{1000}$ $\frac{7}{10}$ $\frac{9}{100}$ $\frac{9}{10}$ $\frac{71}{1000}$ $\frac{7}{100}$ $\frac{9}{1000}$

2 $\frac{3}{4}$ $\frac{13}{25}$ $\frac{7}{50}$ $\frac{19}{20}$ $\frac{1}{2}$ $\frac{13}{50}$ $\frac{7}{20}$ $\frac{11}{25}$ $\frac{17}{20}$ $\frac{1}{4}$

3 $4\frac{19}{1000}$ $6\frac{3}{25}$ $8\frac{9}{20}$ $1\frac{7}{10}$ $12\frac{3}{100}$ $10\frac{63}{1000}$ $2\frac{17}{25}$ $5\frac{3}{20}$

B How many thousandths in the following?

1   0·136   0·075   1·204   0·007   4·071   5·001   0·016   0·009

2   Put in a decimal point to make the <u>7</u> in each number worth 7 hundredths.

     513 467   436 715   643 571   367 154   317 645

3   What is the <u>6</u> worth in each of the **answers** to question **2**?

4   Put in a decimal point to make the <u>5</u> in each number worth 5 thousandths.

     132 415   32 651   96 205   200 531   25

5   What is the <u>2</u> worth in each of the **answers** to question **4**?

C Write out putting in the missing denominators.

1 $7·38 = 7 + \frac{3}{\ } + \frac{8}{\ }$ $6·027 = 6 + \frac{2}{\ } + \frac{7}{\ }$

      $12·375 = 12 + \frac{3}{\ } + \frac{7}{\ } + \frac{5}{\ }$ $7·102 = 7 + \frac{1}{\ } + \frac{2}{\ }$

2   Complete these statements with > or <.

     0·18   0·6   2·017   2·008   4·17   0·417

     0·03   0·027   0·167   0·2   0·076   0·009

     1·06   1·057   46·167   46·176   0·6   0·36

Decimals

A $124 + 3 \cdot 741 + 15 \cdot 077$ $41 \cdot 36 + 1 \cdot 927 + 5\frac{6}{25}$

$4 \cdot 275 + 12\frac{7}{10} + 0 \cdot 93$ $1471 \cdot 6 + 73 \cdot 921 + 5 \cdot 76$

$6\frac{7}{20} + 5 \cdot 21 + 16 \cdot 923$ $0 \cdot 735 + 4 \cdot 88 + 36 \cdot 743$

$12\frac{1}{2} + 16 + 7 \cdot 2 + 12 \cdot 47$ $15\frac{7}{10} + 6\frac{3}{20} + 153 \cdot 754$

$321 \cdot 5 + 0 \cdot 935 + 6 \cdot 74$ $16 \cdot 073 + 8 \cdot 946 + 101 \cdot 52$

B $144 \cdot 337 - 93 \cdot 946$ $132 \cdot 045 - 86 \cdot 925$

$2451 \cdot 96 - 7 \cdot 815$ $53\ 416 \cdot 1 - 7262 \cdot 47$

$24\ 560 \cdot 6 - 43 \cdot 257$ $926 \cdot 04 - 475 \cdot 396$

$27 \cdot 679 - 8 \cdot 489$ $3611 \cdot 3 - 594 \cdot 009$

$6271 - 476 \cdot 998$ $20\ 100 \cdot 63 - 5714 \cdot 29$

C **1** $14 \cdot 136 \times 9$ $321 \cdot 43 \times 8$ $4 \cdot 017 \times 12$ $188 \cdot 06 \times 7$

$0 \cdot 936 \times 11$ $43 \cdot 275 \times 6$ $923 \cdot 6 \times 3$ $149 \cdot 276 \times 4$

  2 $4321 \cdot 06 \times 5$ $1193 \cdot 51 \times 2$ $6251 \cdot 2 \times 12$ $5841 \cdot 031 \times 5$

$20\ 271 \cdot 6 \times 11$ $32\ 151 \cdot 27 \times 4$ $9934 \cdot 007 \times 9$ $2116 \cdot 27 \times 8$

$3313 \cdot 416 \times 6$ $22\ 341 \cdot 6 \times 7$ $5675 \cdot 26 \times 3$ $9354 \cdot 9 \times 2$

Decimals

A

0·56	0·73	0·61	0·39	0·79	0·86
×16	×41	×32	×29	×19	×36

0·19	0·28	0·47	0·92	0·66	0·83
×25	×37	×23	×18	×47	×18

B

1·72	15·45	23·09	2·66	5·27	16·73
×22	×21	×14	×25	×27	×15

13·25	7·45	11·92	21·47	9·32	12·09
×17	×28	×13	×18	×16	×19

C

4·27	11·43	6·21	9·46	14·66
×117	×103	×141	×212	×331

5·21	6·32	12·41	13·09	46·21
×106	×151	×111	×206	×231

D

6) 45·6   9) 80·1   5) 27·5   12) 46·8   11) 68·2

5) 3·15   8) 7·68   7) 3·29   11) 10·34   3) 1·68

7) 93·87   8) 88·72   9) 113·22   4) 64·84   6) 78·12

3) 28·413   5) 23·035   4) 84·108   6) 38·838   11) 26·796

Decimals

A   21 ⟌ 2·646   13 ⟌ 1·014   18 ⟌ 4·356   14 ⟌ 7·392   19 ⟌ 8·949

     15 ⟌ 6·255   17 ⟌ 5·151   16 ⟌ 1·968   23 ⟌ 3·243   24 ⟌ 5·136

B   23 ⟌ 54·05   18 ⟌ 62·28   17 ⟌ 86·19   25 ⟌ 158·50   31 ⟌ 59·21

     32 ⟌ 98·24   27 ⟌ 124·47   19 ⟌ 115·52   41 ⟌ 61·09   62 ⟌ 88·04

C   23 ⟌ 19·366   27 ⟌ 18·954   21 ⟌ 12·033   18 ⟌ 11·538   22 ⟌ 20·152

     34 ⟌ 23·902   27 ⟌ 11·745   36 ⟌ 11·052   24 ⟌ 20·016   33 ⟌ 17·721

D

| 0·57 | 0·61 | 0·93 | 0·47 | 0·82 |
| ×0·7 | ×0·9 | ×0·2 | ×0·6 | ×0·4 |

| 0·98 | 0·34 | 0·54 | 0·66 | 0·39 |
| ×0·8 | ×0·6 | ×0·5 | ×0·3 | ×0·7 |

E

| 12·45 | 15·26 | 22·19 | 32·46 | 27·35 |
| ×1·7 | ×2·7 | ×1·4 | ×2·9 | ×3·2 |

F

| 5·41 | 16·27 | 2·35 | 19·25 | 21·27 |
| ×13·5 | ×1·02 | ×14·3 | ×2·14 | ×8·71 |

G   1·27 × 0·9   32·41 × 3·6   7·49 × 2·41

     11·41 × 13·2   0·96 × 0·17   10·46 × 3·09

Decimals

A   0·8 $\overline{)0·52}$ 0·4 $\overline{)0·14}$ 0·11 $\overline{)0·0792}$ 0·3 $\overline{)0·177}$ 0·5 $\overline{)0·135}$

     0·5 $\overline{)0·17}$ 0·12 $\overline{)0·0156}$ 0·7 $\overline{)0·294}$ 0·6 $\overline{)0·042}$ 0·9 $\overline{)0·612}$

B   0·5 $\overline{)2·630}$ 0·8 $\overline{)3·8}$ 0·12 $\overline{)0·7236}$ 0·6 $\overline{)5·562}$ 0·3 $\overline{)1·023}$

     0·7 $\overline{)2·086}$ 0·11 $\overline{)0·7799}$ 0·9 $\overline{)7·623}$ 0·4 $\overline{)2·50}$ 0·12 $\overline{)0·6408}$

C   8 $\overline{)28·0}$ 5 $\overline{)37·0}$ 6 $\overline{)39·0}$ 12 $\overline{)102·0}$ 4 $\overline{)50·0}$ 5 $\overline{)49·0}$

D   6 $\overline{)44·10}$ 4 $\overline{)22·60}$ 6 $\overline{)43·50}$ 12 $\overline{)33·00}$ 2 $\overline{)19·90}$ 8 $\overline{)44·40}$

E   8 $\overline{)49·00}$ 4 $\overline{)21·50}$ 6 $\overline{)57·75}$ 12 $\overline{)28·50}$ 2 $\overline{)13·75}$ 4 $\overline{)34·50}$

F   35 $\overline{)189·0}$ 18 $\overline{)117·0}$ 14 $\overline{)203·0}$ 15 $\overline{)147·0}$ 25 $\overline{)330·0}$

G   15 $\overline{)137·10}$ 14 $\overline{)87·50}$ 25 $\overline{)184·00}$ 23 $\overline{)132·25}$ 45 $\overline{)177·30}$

H   24 $\overline{)51·00}$ 35 $\overline{)188·16}$ 15 $\overline{)114·42}$ 36 $\overline{)355·5}$ 25 $\overline{)185·6}$

I

99·770 ÷ 22	238·7 ÷ 31	425·10 ÷ 26
117·6 ÷ 16	109·82 ÷ 34	370·00 ÷ 25
201·0 ÷ 15	47·355 ÷ 33	62·028 ÷ 18
196 ÷ 16	181·50 ÷ 25	690·90 ÷ 35
127·26 ÷ 21	181·35 ÷ 45	103·73 ÷ 23

J

260·78 ÷ 118	31·92 ÷ 105	11 087·6 ÷ 212
1044·0 ÷ 144	38·192 ÷ 112	1295·82 ÷ 207

Decimals

A $0.4\,|\,5{\cdot}096$ $0.8\,|\,11{\cdot}400$ $0.9\,|\,19{\cdot}089$ $0.7\,|\,23{\cdot}863$ $0.6\,|\,14{\cdot}130$

   $0.11\,|\,1{\cdot}9877$ $0.12\,|\,3{\cdot}8736$ $0.5\,|\,1{\cdot}2530$ $0.04\,|\,0{\cdot}6940$ $0.3\,|\,3.921$

B $1{\cdot}3\,|\,5{\cdot}473$ $2{\cdot}3\,|\,6{\cdot}187$ $0{\cdot}16\,|\,0{\cdot}8464$ $2{\cdot}5\,|\,67{\cdot}775$ $0{\cdot}35\,|\,8{\cdot}3790$

   $2{\cdot}4\,|\,34{\cdot}128$ $0{\cdot}15\,|\,0{\cdot}9360$ $1{\cdot}9\,|\,95{\cdot}57$ $0{\cdot}32\,|\,0{\cdot}6624$ $1{\cdot}8\,|\,2{\cdot}646$

C $1{\cdot}3\,|\,10{\cdot}673$ $1{\cdot}4\,|\,13{\cdot}230$ $0{\cdot}15\,|\,1{\cdot}0980$ $2{\cdot}1\,|\,10{\cdot}626$ $0{\cdot}26\,|\,1{\cdot}1102$

   $1{\cdot}7\,|\,10{\cdot}778$ $3{\cdot}4\,|\,13{\cdot}090$ $2{\cdot}2\,|\,11{\cdot}550$ $0{\cdot}27\,|\,1{\cdot}1502$ $3{\cdot}1\,|\,19{\cdot}437$

D $11{\cdot}2\,|\,58{\cdot}464$ $1{\cdot}04\,|\,4{\cdot}9192$ $15{\cdot}1\,|\,91{\cdot}355$ $1{\cdot}16\,|\,2{\cdot}3896$ $21{\cdot}3\,|\,72{\cdot}846$

   $20{\cdot}5\,|\,22{\cdot}140$ $0{\cdot}141\,|\,0{\cdot}16074$ $13{\cdot}2\,|\,26{\cdot}928$ $2{\cdot}08\,|\,3{\cdot}2240$ $16{\cdot}1\,|\,34{\cdot}776$

E Divide:

5·130 by 1·5   17·028 by 3·6

18·846 by 0·27   8·7464 by 1·16

5·6945 by 0·35   31·570 by 2·2

4·2140 by 0·35   47·7963 by 2·07

15·906 by 0·22   92·923 by 4·3

1·4464 by 0·32   1·4028 by 0·21

Decimals

A   Write these quantities correct to the: **a** third   **b** second   **c** first decimal place.

2·1274   3·9681   4·0127   5·3482   6·7296

10·1206   9·0079   6·2396   16·2574   8·3521

B   Work out these answers correct to one decimal place.

7 | 4·49   5 | 15·75   4 | 5·3   6 | 25·61   8 | 1·7

0·3 | 0·389   0·9 | 6·5   0·2 | 0·165   0·11 | 0·1677   0·12 | 0·223

1·4 | 0·465   0·26 | 0·8934   2·3 | 3·29   0·17 | 1·399   1·3 | 6·067

C   Work out these answers correct to two decimal places.

7 | 2·99   8 | 8·846   9 | 18·941   12 | 7·51   11 | 46·377

0·2 | 0·1821   0·4 | 0·2905   0·6 | 1·616   0·3 | 1·418   0·5 | 1·9267

0·27 | 0·6062   0·26 | 0·37154   2·1 | 4·337   2·4 | 4·3292

D   Work out these answers correct to three decimal places.

12 | 4·0999   8 | 3·6207   7 | 1·905   11 | 4·379

0·7 | 3·30518   0·6 | 1·5617   0·3 | 10·264   0·4 | 5·2126

1·3 | 0·81529   2·9 | 7·3458   0·47 | 2·02135   1·8 | 1·96295

E   Work out each of these answers correct to: **a** three decimal places **b** two decimal places **c** one decimal place.

0·2148 ÷ 0·13   1·6277 ÷ 2·3   3·47115 ÷ 1·7

8·3291 ÷ 0·37   0·413156 ÷ 0·15   2·7119 ÷ 2·6

Decimal and number problems

1   An athlete throws a javelin 52·74 m, 55·36 m, 51·99 m, 55·63 m.
     What was his average throw?

2

	Jones	Smith	Brown	Gray	Young	Ahmed
1st dive	52·31	51·62	52·53	50·29	54·25	49·29
2nd dive	50·43	53·11	50·23	54·35	51·06	54·92
3rd dive	49·66	49·28	54·55	51·64	50·65	53·55
4th dive	54·04	52·39	48·65	52·32	49·36	48·36

The table shows the marks scored by 6 divers in a diving competition.
The marks are added together after each dive. Who was leading after:

a the 1st dive?

b the 2nd dive?

c the 3rd dive?

d the 4th dive?

Which diver had the highest average mark?

3   A farmer plants 125 rows of cabbage with 147 in each row. How many
     cabbages altogether?

Decimal and number problems

1   There are six turnstiles for the paddock of a football ground. 2475 people went through the 1st turnstile, 1472 the second, 1062 the third, 2025 the fourth, 1001 the fifth, and 2216 the sixth.

    a   How many people were in the paddock?

    b   If the cost was 95p, how much money was taken?

2   A lorry can carry 2475 boxes of eggs. How many boxes are carried by:

    a 48 lorries?   **b** 65 lorries?   **c** 55 lorries?

    If one egg box holds 144 eggs, how many eggs can one lorry carry?

3   A turnstile at a museum shows 36 431 when the museum opens and 38 279 when it closes. How many people have visited the museum?

4   A lottery win of £15,750·63 is won by 23 workers. What is each worker's share?

5   Find the product of:

    a 2749 and 79   **b** 6758·73 and 85

6   Find the quotient of:

    a 56·032 and 1·7   **b** 336·4376 and 1·93

7   Divide the sum of 274·21 and 4326·83 by the difference between 124·02 and 123·78.

Equations

A $84 \div (4 \times 3)$ $=$ $(7 \times 12) + 5$ $=$

$(24 \div 6) + (5 \times 6)$ $=$ $(63 \div 7) - 4$ $=$

$(9 \times 7) - 8$ $=$ $(15 \times 5) - (96 \div 12)$ $=$

$(7 \times 5) + (4 \times 3)$ $=$ $(27 \div 3) + (7 \times 11)$ $=$

$(121 \div 11) - (2 \times 4)$ $=$ $(99 \div 11) - (45 \div 9)$ $=$

B $76 + 35 = 82 + \boxed{}$ $131 + 65 = 96 + \boxed{}$

$129 - 50 = 85 - \boxed{}$ $150 - 73 = 49 + \boxed{}$

$\boxed{} + 16 = 35 + 12$ $\boxed{} + 72 = 96 - 12$

$150 - \boxed{} = 75 + 11$ $67 - \boxed{} = 29 + 14$

$66 - \boxed{} = 29 + 12$ $43 + \boxed{} = 56 - 6$

C $\frac{35}{5} + 9 =$ $125 - \frac{72}{6} =$

$38 - 17 \times 2 =$ $25 \div 5 + 3 =$

$\frac{12}{3} + \frac{36}{6} - 7 =$ $\frac{12}{2} + 7 + \frac{15}{3} =$

$2 \times 7 + 16 =$ $6 \times 9 + 4 \times 5 =$

$3 \times 9 + \frac{16}{4} =$ $\frac{15}{5} + \frac{72}{9} - \frac{12}{4} =$

D $\frac{27}{\boxed{}} + 7 = \frac{20}{4} + 11$ $6 \times \boxed{} + 2 \times 6 = 20 + \frac{40}{4}$

$\frac{\boxed{}}{8} + 3 \times 3 = 2 \times 3 + \frac{35}{5}$ $72 - \frac{45}{5} = 7 \times 8 + \boxed{}$

$6 \times 8 - 4 = \boxed{} + \frac{48}{2}$ $6 \times 7 + 4 \times 9 = 4 \times 7 + \boxed{}$

$12 \times 12 - 63 = 9 \times 9 + \boxed{}$ $\frac{\boxed{}}{5} + 3 \times 4 = 7 \times 4 + 8$

$10 \times 10 - \frac{36}{9} = \boxed{} - \frac{45}{9}$ $7 \times \boxed{} + 5 = 25 + 2 \times 11$

Expressions and equations

A $x = 3$ $y = 7$ $z = 9$ Give the value of:

xy	xyz	$2x + 3$	$4z$	$2xy + 3xz$
$x + 5z$	$xy + yz$	$5y + 2z$	$x + y + 2z$	$6y + 3x$

B Find the value of x.

$3x + 1 = 10$	$5x + 3 = 28$	$2x - 2 = 6$	$x + 9 = 15$
$7x + 4 = 74$	$2x + 12 = 36$	$3x - 5 = 10$	$12x + 10 = 70$

C Write the expression for each statement.

6 less than x	4 times x	9 more than y	$\frac{1}{2}$ of m

r decreased by 6	$\frac{1}{10}$ of 3 times h	x more than 5

twice n plus r	$\frac{1}{4}p$ minus t	15 less than v

D **1** Write the equation for each statement.

15 minus 3 times r equals 3 x divided by 3 is 12 $\frac{1}{8}$ of k is equal to 9

7 plus x equals twice 5   15 less than twice x equals 3   4 times x is 24

2 Solve the equations you have just written.

E Solve:

$(6 \times 125) = (6 \times a) + (6 \times 120)$ $(4 \times 5) + (6 \times 5) = 10 \times x$

$(2 + 5) \times 4 = (2 \times y) + (5 \times y)$ $7 \times 96 = (q \times 96) + (q \times 0)$

$(2 \times 9) = (2 \times 3) + (2 \times n)$ $(36 \times 12) + (n \times 12) = 60 \times 12$

$(m \times 4) + (4 \times 4) = 48$ $30 \times 65 = (v \times 30) + (v \times 35)$

$7 \times (5 + 4) = x + 28$ $m \times 15 = (20 \times 15) + (10 \times 15)$

$5 \times (n + 7) = 25 + 35$ $6 \times 132 = (6 \times b) + (6 \times 100) + (6 \times 30)$

Roman numerals

A Write in our numerals.

1	XXV	IX	XXIII	XXVII	XIX	XXIV
2	XL	LXXXI	LXXVII	XLIV	LXXXIV	XLIII
3	CCC	CXCIV	CCLXXIX	CXLIX	CIV	CXCII
	CXC	CCXVII	CCXC	CXXV	CIX	CCXXX
4	DCCCL	CDIX	DXXXIX	CDLXXXVII		DCLIII
	DXC	DCLXI	DLII	CDXXIV		CDLIV
5	MDCC	MDCCCLIV	MCCCXXI		MCXLIV	
	MDCCXLV	MDCII	MCMLVII		MDCCCLII	

B Write in Roman numerals.

1	39	24	16	7	21	34	19	8
2	47	73	85	64	58	51	69	42
3	341	96	321	252	176	154	329	91
4	479	826	550	720	635	450	890	790
5	1947	1855	1432	1121	1840
	1310	1705	1260	1502	1385
	1688	1468	1940	1892
6	43	96	1141	902	58	520	694
7	354	1072	17	32	1220	87	143
8	2250	1602	25	1111	2222	707	431

Bases – base 10

A Write in full in: **a** figures **b** words.

   1 10^4 10^6 10^2 10^5 10^7 10^8

   2 $10^2 \times 5$ $10^3 \times 5$ $10^5 \times 8$ $10^4 \times 9$ $10^6 \times 3$

B Write in full in figures.

   1 367×10^2 $432{\cdot}5 \times 10^4$ $41{\cdot}29 \times 10^3$ $457{\cdot}6 \times 10^2$

       $11{\cdot}141 \times 10^5$ 2021×10^2 $6{\cdot}23 \times 10^4$ $0{\cdot}359 \times 10^5$

       2164×10^1 $5{\cdot}216 \times 10^3$ $90{\cdot}09 \times 10^2$ $8{\cdot}216 \times 10^4$

   2 $20{\cdot}47 \div 10^1$ $1\,141\,126 \div 10^3$ $173\,000\,000 \div 10^6$ $32{\cdot}7 \div 10^2$

       $2453 \div 10^2$ $16\,000 \div 10^4$ $473{\cdot}1 \div 10^2$ $66\,000 \div 10^2$

       $43\,200 \div 10^5$ $32\,150 \div 10^3$ $1\,200\,000 \div 10^6$ $3941 \div 10^3$

C **1** Complete the table.

	10^6	10^5	10^4	10^3	10^2	10^1	1
3541×10			3	5	4	1	0
4561×1000							
$4{\cdot}263 \times 1\,000\,000$							
$4\,732\,140 \div 10$							
$1\,150\,000 \div 10\,000$							
$6\,300\,000 \div 100\,000$							

D Write in full in figures.

   1 $10^3 \times 10^2$ $10^4 \times 10^2$ $10^6 \times 10^1$ $10^2 \times 10^2$

   2 $10^7 \div 10^2$ $10^9 \div 10^5$ $10^{10} \div 10^6$ $10^5 \div 10^3$

Bases – base 2

A 1 What is the value of:

2^1? 2^3? 2^6? 2^7? 2^5? 2^4? 2^2? 2^8?

2 Write as powers of 2.

16   128   1024   32   4   512   8   2   256   64

B 1 Write the value in denary numbers.

$2^3 \times 2^2$ $2^2 \times 2^4$ $2^3 \times 2^6$ $2^1 \times 2^5$ $2^3 \times 2^4$

2 $2^6 \div 2^4$ $2^5 \div 2^3$ $2^9 \div 2^6$ $2^8 \div 2^5$ $2^8 \div 2^7$

C Base 2

32	16	8	4	2	1	binary number	denary number
2^5	2^4	2^3	2^2	2^1	2^0		
		.		.	.		
.	.		.	.	.		
.		.	.		.		
.		.		.	.		
	.	.		.	.		
.	.			.	.		

1 Write down: **a** the binary numbers **b** the denary numbers, represented by the dots on the chart.

2 Write these numbers as denary numbers.

11011_2 101010_2 111_2 1001_2 10001_2

D 1 Write as binary numbers.

63_{10} 47_{10} 29_{10} 35_{10} 18_{10} 7_{10}

2
$$10101_2$$
$$1011_2$$
$$+1110_2$$

$$1111_2$$
$$1010_2$$
$$+101_2$$

$$11110_2$$
$$10001_2$$
$$+1011_2$$

$$111_2$$
$$1010_2$$
$$+11111_2$$

$$10010_2$$
$$11111_2$$
$$+1001_2$$

3
$$1001_2$$
$$-110_2$$

$$11101_2$$
$$-1001_2$$

$$10010_2$$
$$-1001_2$$

$$11110_2$$
$$-101_2$$

$$101010_2$$
$$11110_2$$

Averages

A Find the average of each set of numbers.

1	3·24	4·60	2·71	9·81		
2	21·21	9·21	4·30	2·27	6·16	
3	6·26	19·46	23·75	13·44	12·71	19·54
4	2745	5621	4220	3174		
5	227 432	347 609	476 540			

B

1 The average weight of nine boys was 38·5 kg. What was their total weight?

2 The average cost of seven articles was £2·25. What was the total cost?

3 On a touring holiday which lasted 14 days, the average distance covered each day was 48 km. What was the total distance travelled?

4 A milkman delivered 674 l of milk each day. If he made no deliveries on Sunday, how much milk did he deliver in one week?

5 The average height of six boys was 152 cm. The total height of five of the boys was 765 cm, so how tall was the other boy?

6 Eight parcels weigh 760 g. The weights of seven of the parcels are 93 g, 87 g, 91 g, 96 g, 98 g, 89 g, 90 g.

  a What does the other parcel weigh?

  b How much more than the average weight does this parcel weigh?

7 The average spending money of five children is £5·32. The spending money of four of the children is £4·80, £4·65, £3·95 and £6·20. Find:

  a the total spending money

  b the spending money of the other child.

Money

A **1** Give the change you would receive from: **a** a fifty pence
b a pound if you spent:

23p	£0·47	38p	£0·27	41p	£0·12
£0·16	48p	35p	£0·33	9p	£0·25

2 Give the total of each of these six amounts of money.

	£5	£1	50p	20p	10p	5p	2p	1p	total
a	2	1	5	6		8	7	3	
b	1	9	3	12	7	5	8	1	
c		12	7	5	4	6	5		
d	5			21	5	4	6	2	
e	3	4		7	9		12	6	
f	2	5	4	7	8	10	2	3	

B   £8·12 + £0·17 + £4·47   £145·26 + £73·88

£24·76 + £18·07 + £0·99   £134·14 + £26·99

£16·43 + £21·16 + £34·34   £232·67 + £141·75

£23·16 + £15·27 + £47·43   £134·16 + £362·15

£19·04 + £16·12 + £0·79   £204·19 + £96·82

C   £73·16 − £47·88   £200 − £147·16

£146·08 − £79·76   £152·20 − £123·18

£104·26 − £93·69   £140·36 − £73·49

£147·32 − £66·45   £194·07 − £189·22

£99·00 − £43·43   £250 − £236·86

Money

A   1

no	item	cost each	total
7	tin of beans	51p	
3	tin of meat	97p	
6	loaf	92p	
5	bag of sugar	53p	
8	packet of tea	46p	
9	jelly	17p	
4	packet of cereals	72p	

Copy and complete the table.

2   What is:

$\frac{1}{6}$ of   24p   96p   30p   36p   12p   54p   42p?

$\frac{1}{4}$ of   24p   52p   44p   36p   28p   32p   48p?

$\frac{1}{3}$ of   39p   33p   21p   36p   27p   51p   24p?

$\frac{1}{8}$ of   32p   40p   48p   56p   64p   80p   72p?

B

£6·17 × 7	£25·06 × 9	£0·87 × 12
£18·42 × 6	£23·27 × 11	£41·36 × 8
£0·17 × 15	£1·16 × 23	£2·19 × 21
£32·41 × 25	£6·24 × 18	£16·48 × 19

C

£74·40 ÷ 12	£134·47 ÷ 7	£45·81 ÷ 9
£46·53 ÷ 11	£90·24 ÷ 8	£100·86 ÷ 6
£54·99 ÷ 13	£42·49 ÷ 7	£20·25 ÷ 15
£30·38 ÷ 14	£94·76 ÷ 23	£3·74 ÷ 17

D   Multiply $\frac{1}{5}$ of £2·70 by 6.

Divide 7 times £2·31 by 3.

Costing

1 Find the cost of:

500 g at 96p per kg	$2\frac{1}{2}$ kg at 76p per $\frac{1}{2}$ kg
400 ml at £1·20 per litre	50 cm at 62p per metre
6 m at 27p per metre	$5\frac{1}{2}$ l at 13p per $\frac{1}{4}$ litre
700 g at 30p per kg	700 mm at 70p per metre
250 ml at 48p per litre	6·5 kg at 13p per 0·5 kg

2 Copy and complete the table.

1 kg	500 g	200 g	100 g
£1·00			
50p			
40p			
30p			
20p			

Find the cost of:

700 g at 50p per kg	300 g at 40p per kg
800 g at 30p per kg	300 g at 20p per kg
600 g at 20p per kg	600 g at 30p per kg

3 Find the cost of:

500 g at 60p per kg	400 g at 50p per kg	250 g at £1·20 per kg
100 g at 90p per kg	500 g at £1·40 per kg	700 g at 70p per kg

4 Pork costs £3·70 per $\frac{1}{2}$ kg. Find the cost of:

100 g   200 g   50 g   500 g   450 g   350 g   950 g   550 g

5 $3\frac{1}{2}$ metres of cloth costs £3·50. Find the cost of:

$\frac{1}{2}$ m   1 m $4\frac{1}{2}$ m $6\frac{1}{2}$ m   3 m $2\frac{1}{2}$ m   7 m   9 m

6 Envelopes cost £2·40 for 50. Find the cost of:

100   25   200   400   350   750   225   275   375   425

7 $1\frac{1}{4}$ litres of orange squash costs 80p. How much for:

$\frac{1}{4}$ l $\frac{1}{2}$ l $\frac{3}{4}$ l   1 l $2\frac{1}{2}$ l $3\frac{3}{4}$ l $4\frac{1}{4}$ l   5 l $5\frac{1}{2}$ l $6\frac{1}{4}$ l

Profit and loss

A **1** Copy and complete.

cost price	£3·25	£5·50	£8·20	£10·50	£15·60	£18·50
selling price	£4·00	£6·99	£9·35	£11·49	£17·39	£19·99
profit						

2 Copy and complete the table showing the sale of damaged goods.

cost price	£7·50	£9·50	£12·50	£15·25	£17·75	£20·50
selling price	£6·25	£7·45	£10·75	£13·80	£16·95	£14·65
loss						

B **1** A car was bought for £2,575. It was sold later for £1,950. What was the loss?

2 What must a £3·50 box of apples be sold at to make a profit of 85p?

3 A shopkeeper buys 400 lettuce at 11p each. How much profit will he make if he sells them at 14p each?

4 A dealer pays £24·50, £32·60, £29·80, and £16·50 for four vases. He sells the vases for £130·00 altogether. What is his average profit per vase?

5 Fourteen pairs of ladders each cost £7·25. How much must each pair be sold for, to make a total profit of £35·00?

6 15 lamps were bought for £138·75. **a** How much each did they cost? **b** If each lamp was sold for £12·50, what was the total profit?

7 Four shop-soiled rugs which cost £262 altogether, are sold at £49·95 each. What is the amount of money lost?

8 How much would be saved by buying a car for £6,750 cash instead of paying 24 payments of £325·75?

Fractions

A What fraction or part of these is shaded?

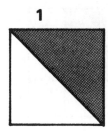
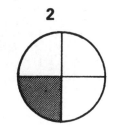
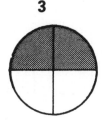
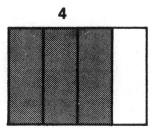

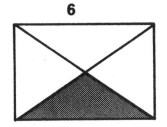
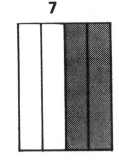
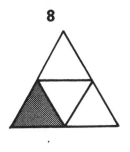

B Which of these lines are divided into halves?

C Which of these lines are divided into quarters?

Fractions

A   Express in lowest terms: **a** the shaded part   **b** the unshaded part.

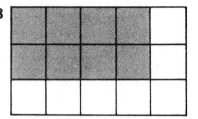

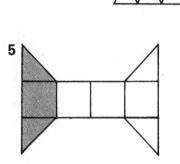

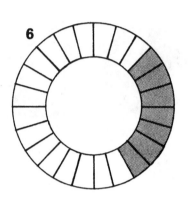

B   How many:

halves	$2\frac{1}{2}$	$3\frac{1}{2}$	$4\frac{1}{2}$	5	6	$9\frac{1}{2}$	7	8	?
tenths	$2\frac{1}{5}$	$3\frac{3}{10}$	$4\frac{3}{5}$	$2\frac{1}{10}$	$6\frac{7}{10}$	$3\frac{4}{5}$	$1\frac{9}{10}$	$2\frac{2}{5}$	?
sixteenths	$1\frac{1}{16}$	$2\frac{1}{8}$	$3\frac{11}{16}$	$1\frac{15}{16}$	$3\frac{5}{8}$	$3\frac{7}{8}$	$2\frac{9}{16}$	$4\frac{5}{16}$	?
fifths	$2\frac{1}{5}$	$6\frac{3}{5}$	$3\frac{1}{5}$	$2\frac{2}{5}$	$4\frac{3}{5}$	$6\frac{3}{5}$	$2\frac{4}{5}$	$5\frac{4}{5}$	?

C   Complete the series with fractions in lowest terms.

$\frac{1}{2}$ $\frac{7}{12}$ ☐ $\frac{3}{4}$ ☐ $\frac{11}{12}$ $2\frac{14}{15}$ $2\frac{13}{15}$ ☐ $2\frac{11}{15}$ ☐   ☐

$\frac{1}{4}$ $\frac{3}{10}$ $\frac{7}{20}$ ☐ $\frac{9}{20}$ ☐ $6\frac{19}{24}$ $6\frac{3}{4}$ $6\frac{17}{24}$ ☐ $6\frac{5}{8}$ ☐   ☐

$1\frac{1}{3}$ $1\frac{1}{2}$ ☐ $1\frac{5}{6}$ ☐ $\frac{13}{16}$ $\frac{3}{4}$ $\frac{11}{16}$ ☐ $\frac{9}{16}$ $\frac{1}{2}$ ☐   ☐

D   Arrange in order of size – smallest first.

$3\frac{4}{5}$ $3\frac{13}{20}$ $3\frac{7}{10}$ $3\frac{3}{4}$ $6\frac{7}{24}$ $6\frac{1}{3}$ $6\frac{3}{8}$ $6\frac{1}{4}$

$1\frac{3}{4}$ $1\frac{5}{8}$ $1\frac{9}{16}$ $1\frac{15}{32}$ $4\frac{7}{15}$ $4\frac{3}{10}$ $4\frac{2}{5}$ $4\frac{9}{20}$

$3\frac{2}{3}$ $3\frac{5}{6}$ $3\frac{4}{9}$ $3\frac{7}{12}$ $2\frac{23}{24}$ $2\frac{11}{18}$ $2\frac{7}{12}$ $2\frac{5}{6}$

Fractions – addition and subtraction

A **1** Give the lowest common multiple for:

4 and 8	3 and 6	5 and 10	10 and 20	8 and 16
2 and 5	4 and 5	3 and 8	2 and 9	3 and 7
5 and 6	4 and 6	4 and 3	7 and 5	6 and 8

2 Change to mixed numbers.

$$\frac{15}{4} \qquad \frac{27}{10} \qquad \frac{36}{7} \qquad \frac{41}{8} \qquad \frac{15}{6} \qquad \frac{29}{20} \qquad \frac{20}{3} \qquad \frac{23}{11}$$

B

$\frac{3}{5} + \frac{5}{8}$	$\frac{2}{3} + \frac{7}{12}$	$\frac{9}{24} + \frac{7}{10}$	$\frac{7}{8} + \frac{3}{10}$
$\frac{8}{9} + \frac{1}{6}$	$\frac{13}{25} + \frac{9}{10}$	$\frac{3}{4} + \frac{4}{5}$	$\frac{13}{16} + \frac{5}{8}$
$\frac{13}{20} + \frac{3}{5}$	$\frac{3}{4} + \frac{7}{8}$	$\frac{11}{12} + \frac{5}{8}$	$\frac{9}{11} + \frac{1}{2}$

C

$4\frac{3}{5} + 2\frac{1}{10}$	$2\frac{7}{8} + 3\frac{5}{16}$	$6\frac{7}{20} + 4\frac{7}{10}$	$1\frac{5}{8} + \frac{11}{16}$
$\frac{19}{20} + 1\frac{3}{4}$	$3\frac{5}{6} + 4\frac{11}{12}$	$2\frac{19}{25} + 4\frac{4}{5}$	$5\frac{5}{18} + 2\frac{2}{3}$
$\frac{19}{24} + 6\frac{1}{6}$	$9\frac{2}{3} + 4\frac{7}{15}$	$2\frac{4}{5} + 8\frac{3}{4}$	$5\frac{9}{22} + 2\frac{1}{2}$

D

$\frac{7}{13} + \frac{19}{26}$	$\frac{4}{5} + \frac{11}{12}$	$\frac{7}{15} + \frac{5}{12}$	$\frac{5}{9} + \frac{4}{5}$
$\frac{5}{6} + \frac{8}{9}$	$\frac{19}{20} + \frac{11}{30}$	$\frac{7}{8} + \frac{8}{9}$	$\frac{7}{15} + \frac{29}{30}$
$\frac{11}{15} + \frac{41}{45}$	$\frac{23}{24} + \frac{19}{36}$	$\frac{5}{7} + \frac{1}{6}$	$\frac{6}{7} + \frac{5}{9}$

E

$3\frac{5}{18} + 2\frac{7}{12}$	$6\frac{5}{7} + 3\frac{5}{14}$	$9\frac{7}{16} + 6\frac{3}{10}$	$2\frac{6}{7} + 5\frac{2}{9}$
$7\frac{1}{13} + 5\frac{4}{39}$	$6\frac{7}{20} + 8\frac{8}{15}$	$2\frac{4}{9} + 3\frac{4}{15}$	$8\frac{5}{33} + 4\frac{7}{11}$
$4\frac{1}{15} + 8\frac{19}{60}$	$2\frac{2}{3} + 1\frac{19}{20}$	$6\frac{4}{7} + 5\frac{3}{4}$	$4\frac{5}{12} + 3\frac{9}{15}$

F $7\frac{7}{8} + 5\frac{13}{20} + 6\frac{4}{5}$ $\qquad\qquad\qquad$ $8\frac{11}{15} + 4\frac{13}{20} + 6\frac{19}{30}$

Fractions – addition and subtraction

A **1** Give the lowest common denominator for:

$\frac{1}{20}$ and $\frac{1}{30}$ $\qquad$ $\frac{1}{25}$ and $\frac{1}{10}$ $\qquad$ $\frac{1}{16}$ and $\frac{1}{24}$ $\qquad$ $\frac{1}{12}$ and $\frac{1}{10}$

$\frac{1}{15}$ and $\frac{1}{12}$ $\qquad$ $\frac{1}{25}$ and $\frac{1}{4}$ $\qquad$ $\frac{1}{14}$ and $\frac{1}{21}$ $\qquad$ $\frac{1}{9}$ and $\frac{1}{8}$

$\frac{1}{20}$ and $\frac{1}{6}$ $\qquad$ $\frac{1}{8}$ and $\frac{1}{5}$ $\qquad$ $\frac{1}{13}$ and $\frac{1}{4}$ $\qquad$ $\frac{1}{7}$ and $\frac{1}{9}$

2 What are the missing numerators?

$\frac{7}{20} = \frac{\Box}{40}$ $\qquad$ $\frac{7}{8} = \frac{\Box}{56}$ $\qquad$ $\frac{2}{9} = \frac{\Box}{63}$ $\qquad$ $\frac{11}{25} = \frac{\Box}{50}$

$\frac{\Box}{20} = \frac{4}{5}$ $\qquad$ $\frac{\Box}{28} = \frac{5}{7}$ $\qquad$ $\frac{\Box}{60} = \frac{9}{15}$ $\qquad$ $\frac{\Box}{15} = \frac{2}{3}$

B $\frac{7}{8} - \frac{3}{4}$ $\qquad$ $\frac{9}{10} - \frac{11}{20}$ $\qquad$ $\frac{5}{6} - \frac{17}{24}$ $\qquad$ $\frac{15}{16} - \frac{3}{4}$

$\frac{9}{11} - \frac{1}{2}$ $\qquad$ $\frac{24}{25} - \frac{4}{5}$ $\qquad$ $\frac{3}{4} - \frac{1}{6}$ $\qquad$ $\frac{7}{12} - \frac{1}{5}$

$\frac{17}{40} - \frac{19}{80}$ $\qquad$ $\frac{6}{7} - \frac{9}{21}$ $\qquad$ $\frac{8}{9} - \frac{15}{18}$ $\qquad$ $\frac{13}{16} - \frac{3}{8}$

C $2\frac{1}{3} - 1\frac{5}{8}$ $\qquad$ $4\frac{3}{4} - 2\frac{9}{10}$ $\qquad$ $6\frac{7}{8} - 3\frac{3}{10}$ $\qquad$ $7\frac{4}{7} - 2\frac{3}{14}$

$8\frac{7}{10} - 4\frac{3}{4}$ $\qquad$ $2\frac{4}{5} - 1\frac{7}{20}$ $\qquad$ $9\frac{19}{24} - 6\frac{1}{3}$ $\qquad$ $5\frac{5}{8} - 3\frac{1}{12}$

$9\frac{4}{25} - 3\frac{11}{50}$ $\qquad$ $4\frac{3}{4} - 3\frac{2}{3}$ $\qquad$ $5\frac{5}{12} - 3\frac{5}{6}$ $\qquad$ $2\frac{9}{10} - 1\frac{1}{2}$

D $\frac{19}{30} - \frac{7}{20}$ $\qquad$ $\frac{6}{7} - \frac{5}{9}$ $\qquad$ $\frac{7}{9} - \frac{8}{15}$ $\qquad$ $\frac{4}{5} - \frac{11}{45}$

$\frac{19}{25} - \frac{35}{50}$ $\qquad$ $\frac{19}{20} - \frac{5}{6}$ $\qquad$ $\frac{17}{18} - \frac{7}{12}$ $\qquad$ $\frac{7}{8} - \frac{6}{7}$

$\frac{6}{13} - \frac{1}{4}$ $\qquad$ $\frac{11}{15} - \frac{5}{12}$ $\qquad$ $\frac{7}{11} - \frac{14}{33}$ $\qquad$ $\frac{9}{14} - \frac{5}{21}$

E $2\frac{7}{9} - 1\frac{9}{10}$ $\qquad$ $9\frac{5}{16} - 6\frac{17}{24}$ $\qquad$ $10\frac{7}{30} - 5\frac{1}{4}$ $\qquad$ $5\frac{11}{15} - 3\frac{23}{25}$

$3\frac{1}{3} - 2\frac{7}{20}$ $\qquad$ $2\frac{11}{12} - 1\frac{7}{10}$ $\qquad$ $7\frac{2}{9} - 4\frac{2}{27}$ $\qquad$ $5\frac{2}{3} - 3\frac{9}{16}$

$5\frac{7}{8} - 1\frac{4}{5}$ $\qquad$ $8\frac{3}{10} - 6\frac{19}{25}$ $\qquad$ $2\frac{11}{16} - 1\frac{5}{12}$ $\qquad$ $4\frac{9}{20} - 2\frac{11}{15}$

F $6\frac{3}{8} + 5\frac{4}{5} - 7\frac{9}{20}$ $\qquad\qquad\qquad$ $8\frac{13}{20} + 4\frac{3}{5} - 6\frac{2}{3}$

$5\frac{3}{4} - 2\frac{7}{12} + 6\frac{3}{10}$ $\qquad\qquad\qquad$ $11\frac{11}{12} - 8\frac{7}{16} + 5\frac{5}{24}$

Fractions – multiplication and division

A **1** Cancel to lowest terms.

$\frac{45}{50}$ $\quad$ $\frac{36}{48}$ $\quad$ $\frac{33}{45}$ $\quad$ $\frac{60}{72}$ $\quad$ $\frac{25}{100}$ $\quad$ $\frac{45}{60}$ $\quad$ $\frac{18}{27}$ $\quad$ $\frac{36}{42}$ $\quad$ $\frac{72}{80}$ $\quad$ $\frac{52}{65}$

2 Change to improper fractions.

$7\frac{3}{4}$ $\quad$ $8\frac{5}{6}$ $\quad$ $9\frac{2}{13}$ $\quad$ $1\frac{7}{12}$ $\quad$ $4\frac{13}{20}$ $\quad$ $2\frac{7}{10}$ $\quad$ $5\frac{11}{16}$ $\quad$ $4\frac{1}{60}$ $\quad$ $2\frac{2}{15}$ $\quad$ $1\frac{7}{30}$

B $\frac{4}{5} \times \frac{15}{20}$ $\qquad$ $\frac{3}{8} \times \frac{24}{25}$ $\qquad$ $\frac{19}{20} \times \frac{6}{7}$ $\qquad$ $\frac{3}{4} \times \frac{8}{9}$

$\frac{11}{12} \times \frac{6}{11}$ $\qquad$ $\frac{15}{16} \times \frac{4}{5}$ $\qquad$ $\frac{13}{20} \times \frac{8}{9}$ $\qquad$ $\frac{21}{25} \times \frac{15}{16}$

$\frac{2}{3} \times \frac{9}{10}$ $\qquad$ $\frac{14}{15} \times \frac{17}{21}$ $\qquad$ $\frac{11}{40} \times \frac{10}{11}$ $\qquad$ $\frac{9}{16} \times \frac{4}{15}$

C $2\frac{1}{5} \times 2\frac{3}{4}$ $\qquad$ $4\frac{1}{4} \times 1\frac{1}{3}$ $\qquad$ $2\frac{4}{5} \times 1\frac{3}{7}$ $\qquad$ $2\frac{1}{10} \times 2\frac{1}{6}$

$5\frac{1}{2} \times 2\frac{2}{5}$ $\qquad$ $7\frac{1}{2} \times 1\frac{3}{10}$ $\qquad$ $6\frac{1}{4} \times 2\frac{2}{15}$ $\qquad$ $3\frac{1}{3} \times 1\frac{1}{20}$

D $\frac{15}{16} \times \frac{24}{25}$ $\qquad$ $\frac{11}{60} \times \frac{48}{55}$ $\qquad$ $\frac{13}{14} \times \frac{28}{39}$ $\qquad$ $\frac{15}{28} \times \frac{7}{10}$

$\frac{9}{56} \times \frac{21}{25}$ $\qquad$ $\frac{3}{49} \times \frac{14}{45}$ $\qquad$ $\frac{13}{18} \times \frac{27}{52}$ $\qquad$ $\frac{35}{36} \times \frac{48}{55}$

E $\frac{5}{12} \times 1\frac{7}{15}$ $\qquad$ $6\frac{2}{3} \times 1\frac{8}{25}$ $\qquad$ $4\frac{2}{7} \times 2\frac{1}{10}$ $\qquad$ $2\frac{13}{18} \times 5\frac{1}{7}$

$2\frac{7}{24} \times 2\frac{2}{11}$ $\qquad$ $1\frac{5}{16} \times 2\frac{2}{21}$ $\qquad$ $3\frac{1}{9} \times 2\frac{1}{4}$ $\qquad$ $1\frac{19}{36} \times 9\frac{3}{5}$

F $\frac{7}{8} \times \frac{4}{5} \times \frac{5}{21}$ $\qquad\qquad\qquad$ $\frac{13}{16} \times \frac{12}{25} \times \frac{15}{26}$

$\frac{2}{3} \times \frac{9}{10} \times \frac{25}{27}$ $\qquad\qquad\qquad$ $\frac{11}{60} \times \frac{24}{33} \times \frac{9}{14}$

G $2\frac{1}{2} \times 3\frac{1}{5} \times 3\frac{3}{4}$ $\qquad\qquad$ $2\frac{2}{9} \times 1\frac{5}{6} \times 1\frac{3}{22}$

$1\frac{5}{16} \times 1\frac{1}{9} \times 4\frac{4}{5}$ $\qquad\qquad$ $5\frac{1}{7} \times 4\frac{2}{3} \times 2\frac{3}{8}$

Fractions – multiplication and division

A Give the highest common factor of:

45 and 60	63 and 21	24 and 36	25 and 35
21 and 28	80 and 100	24 and 32	15 and 18
25 and 75	60 and 48	44 and 66	72 and 84

B Change to improper fractions.

$2\frac{7}{8}$　　$5\frac{13}{14}$　　$4\frac{5}{6}$　　$3\frac{11}{12}$　　$8\frac{13}{20}$　　$2\frac{19}{25}$　　$5\frac{3}{16}$　　$8\frac{7}{10}$

C $\frac{4}{5} \div \frac{8}{15}$　　　　$\frac{3}{10} \div \frac{9}{25}$　　　　$\frac{5}{16} \div \frac{15}{16}$　　　　$\frac{7}{8} \div \frac{14}{15}$

$\frac{11}{12} \div \frac{7}{24}$　　　　$\frac{13}{21} \div \frac{5}{7}$　　　　$\frac{9}{20} \div \frac{6}{25}$　　　　$\frac{8}{35} \div \frac{16}{21}$

$\frac{7}{24} \div \frac{5}{16}$　　　　$\frac{11}{50} \div \frac{22}{25}$　　　　$\frac{35}{36} \div \frac{25}{48}$　　　　$\frac{17}{20} \div \frac{23}{60}$

D $2\frac{1}{2} \div 2\frac{2}{5}$　　　$7\frac{3}{5} \div 10\frac{2}{5}$　　　$6\frac{1}{4} \div 1\frac{3}{7}$　　　$8\frac{1}{3} \div 1\frac{7}{8}$

$12\frac{1}{2} \div 8\frac{3}{4}$　　　$2\frac{1}{7} \div 2\frac{13}{16}$　　　$2\frac{7}{10} \div 2\frac{1}{10}$　　　$1\frac{6}{7} \div 1\frac{3}{4}$

$1\frac{13}{20} \div 4\frac{8}{9}$　　　$2\frac{2}{5} \div 1\frac{5}{11}$　　　$5\frac{5}{8} \div 3\frac{3}{4}$　　　$3\frac{8}{9} \div 2\frac{6}{7}$

E $\frac{14}{25} \div \frac{28}{45}$　　　　$\frac{33}{45} \div \frac{66}{75}$　　　　$\frac{15}{16} \div \frac{73}{80}$　　　　$\frac{18}{19} \div \frac{27}{38}$

$\frac{20}{21} \div \frac{10}{63}$　　　　$\frac{45}{77} \div \frac{20}{33}$　　　　$\frac{75}{77} \div \frac{25}{44}$　　　　$\frac{26}{45} \div \frac{13}{30}$

F $2\frac{22}{25} \div 2\frac{7}{10}$　　　$3\frac{7}{15} \div 2\frac{3}{5}$　　　$\frac{19}{36} \div 4\frac{3}{4}$　　　$7\frac{3}{7} \div 2\frac{8}{9}$

$4\frac{2}{7} \div 3\frac{7}{11}$　　　　$2\frac{2}{15} \div \frac{16}{25}$　　　$8\frac{1}{10} \div 1\frac{19}{35}$　　　$4\frac{8}{11} \div 5\frac{1}{5}$

G $3\frac{3}{4} \times \frac{12}{15} \div 1\frac{4}{5}$　　　　　　　$4\frac{3}{8} \times 1\frac{4}{5} \div 1\frac{5}{16}$

$2\frac{4}{19} \times 4\frac{2}{9} \div 2\frac{5}{8}$　　　　　　$4\frac{1}{11} \times 3\frac{3}{10} \div 2\frac{1}{4}$

H $(1\frac{4}{5} - \frac{7}{10}) \div (5\frac{3}{4} + 2\frac{1}{2})$　　　　$(2\frac{5}{16} - \frac{5}{8}) \div (4\frac{5}{6} - 2\frac{7}{12})$

$(3\frac{1}{3} + 4\frac{5}{6}) \div (3\frac{2}{5} - 1\frac{3}{10})$　　　$(6\frac{2}{5} - 4\frac{3}{8}) \div (2\frac{1}{2} + 1\frac{11}{20})$

Fractions – mixed examples

A **1** Change to vulgar fractions in lowest terms.

0·53	1·17	2·35	4·8	14·325	6·03
11·006	15·75	12·62	19·47	0·015	3·426

2 Change to decimal fractions.

$2\frac{3}{10}$ $4\frac{7}{20}$ $6\frac{9}{25}$ $11\frac{7}{10}$ $13\frac{4}{5}$ $6\frac{17}{50}$ $\frac{19}{1000}$ $1\frac{17}{20}$

$5\frac{3}{8}$ $9\frac{14}{25}$ $7\frac{13}{20}$ $6\frac{5}{8}$ $\frac{37}{1000}$ $2\frac{17}{500}$ $1\frac{31}{250}$ $7\frac{113}{125}$

$2\frac{7}{8}$ $3\frac{13}{1000}$ $4\frac{73}{250}$ $6\frac{97}{125}$ $6\frac{151}{500}$ $1\frac{18}{25}$ $\frac{11}{20}$ $2\frac{479}{500}$

B Change to decimal fractions by dividing.

Example: $\frac{1}{5} = 5\overline{)1\cdot0}$ (quotient $0\cdot2$)

1 $\frac{9}{10}$ $\frac{19}{20}$ $\frac{3}{5}$ $\frac{16}{25}$ $\frac{1}{8}$ $\frac{3}{20}$ $\frac{11}{125}$ $\frac{39}{50}$ $\frac{67}{500}$ $\frac{3}{8}$ $\frac{9}{20}$ $\frac{21}{25}$

2 $4\frac{7}{8}$ $2\frac{3}{10}$ $6\frac{2}{5}$ $8\frac{7}{25}$ $5\frac{1}{20}$ $2\frac{51}{125}$ $4\frac{37}{50}$ $6\frac{431}{500}$

C Change to decimal fractions by dividing. Give your answers correct to 3 decimal places.

1 $\frac{7}{11}$ $\frac{14}{15}$ $\frac{7}{9}$ $\frac{13}{16}$ $\frac{2}{7}$ $\frac{11}{12}$ $\frac{5}{6}$ $\frac{5}{21}$ $\frac{17}{18}$ $\frac{2}{3}$

2 $4\frac{5}{13}$ $2\frac{13}{18}$ $5\frac{3}{14}$ $9\frac{19}{33}$ $6\frac{10}{11}$ $8\frac{13}{15}$ $2\frac{11}{16}$ $1\frac{9}{17}$

D Arrange in order of size — largest first.

1 $\frac{5}{12}$ 0·53 $\frac{5}{8}$ $\frac{5}{11}$ 0·473 $\frac{4}{9}$ 0·432

2 0·251 $\frac{3}{8}$ $\frac{3}{10}$ 0·335 $\frac{4}{15}$ $\frac{5}{18}$ 0·326

3 0·755 $\frac{15}{24}$ $\frac{19}{25}$ 0·695 $\frac{8}{11}$ 0·732 $\frac{13}{16}$

Fractions – problems

A Calculate:

$\frac{7}{10}$ of   **a** £4·35   **b** 12 m   **c** 550   **d** 7·9   **e** 2·75 km

$\frac{5}{8}$ of   **a** 320   **b** 3·48 l   **c** 2 040   **d** 640 g   **e** 780 mm

$\frac{11}{20}$ of   **a** 22 040   **b** £17·60   **c** $2\frac{3}{4}$ kg   **d** 960 m   **e** 23 tonnes

B Find the whole when:

a $\frac{3}{4}$ is 342   **b** $\frac{5}{8}$ is 265 g   **c** $\frac{9}{10}$ is 891 l   **d** $\frac{5}{6}$ is $12\frac{1}{2}$

e $\frac{11}{12}$ is 352p   **f** $\frac{14}{15}$ is 294 m   **g** $\frac{19}{20}$ is 57 km   **h** $\frac{7}{11}$ is $5\frac{1}{4}$

C **1** Emily walks $\frac{4}{5}$ km to school. Ethan walks $\frac{3}{4}$ km to school. How much further does Emily walk?

2 A third of Ryan's money is 75p. What would $\frac{3}{5}$ be?

3 In a collection of stamps, $\frac{7}{20}$ were British, $\frac{1}{10}$ were Commonwealth. The total number of stamps in the collection was 460. How many stamps were:

    a British?   **b** Commonwealth?   **c** other countries?

4 Emily spends $\frac{2}{3}$ of her money on a dress, $\frac{1}{4}$ on a pair of shoes and has £2·50 left. How much money did: **a** she have to begin with?
   b the dress cost?   **c** the shoes cost?

5 In four days a car travels $12\frac{3}{4}$ km, $16\frac{2}{3}$ km, $19\frac{7}{10}$ km and $8\frac{1}{2}$ km. How far did it travel altogether?

6 Take the sum of $4\frac{3}{10}$ and $2\frac{11}{12}$ from the difference between $4\frac{4}{5}$ and $12\frac{7}{10}$.

7 Find the product of: **a** $4\frac{2}{7}$ and $4\frac{9}{10}$ **b** $8\frac{1}{10}$ and $2\frac{7}{9}$

8 Subtract the sum of $4\frac{3}{16}$ and $2\frac{5}{8}$ from the product of $5\frac{1}{7}$ and $1\frac{5}{12}$.

Percentages

A Use the signs >, < or = to complete the following.

8% $\frac{2}{25}$	70% $\frac{4}{5}$	35% $\frac{3}{10}$	55% $\frac{11}{20}$
$\frac{3}{4}$ 65%	$\frac{19}{20}$ 76%	$\frac{17}{25}$ 72%	$\frac{3}{5}$ 62%
$\frac{5}{8}$ $72\frac{1}{2}$%	$\frac{3}{20}$ 12%	$\frac{14}{25}$ 56%	32% $\frac{7}{20}$

B Find the value of:

50% of 90	20% of £2·05	5% of 250 kg
10% of 120	50% of 36	10% of 970
4% of 325	$12\frac{1}{2}$% of 176	5% of 150
25% of 1000	10% of 325 l	2% of 600 g
20% of £3·50	5% of 80p	$12\frac{1}{2}$% of 64
25% of 480	2% of 2500	20% of 105

C Find the value of:

1 a 4%	**b** 12%	**c** 16%	**d** 32%	**e** 48%	of 125
2 a 20%	**b** 40%	**c** 60%	**d** 80%		of £2·50
3 a 10%	**b** 30%	**c** 70%	**d** 90%		of 450
4 a 2%	**b** 6%	**c** 18%	**d** 34%	**e** 46%	of 300 g
5 a 5%	**b** 15%	**c** 35%	**d** 45%	**e** 55%	of £5·00
6 a $12\frac{1}{2}$%	**b** $37\frac{1}{2}$%	**c** $62\frac{1}{2}$%	**d** $87\frac{1}{2}$%		of 1000

D

10% of 50p =	70% of £3·50 =	45% of 330 m =
40% of 75 g =	25% of 84 kg =	26% of 150 t =
75% of 450 =	50% of £100 =	90% of 450p =
30% of 960p =	$12\frac{1}{2}$% of 72 =	35% of 80 =
12% of 175 cm =	16% of 50 =	80% of £625 =
60% of £5·00 =	20% of 550 l =	15% of 120 =

Percentages

1 Increase these numbers by: **a** 10%   **b** 25%.

| 490 | 750 | 680 | 484 | 768 | 924 |

2 Decrease each number by: **a** 20%   **b** 5%.

| 3754 | 8902 | 7426 | 6891 | 5249 | 7348 |

3

items	sale price	a	b
socks	95p		
shirts	£2·95		
shoes	£5·60		
ties	£1·25		
coats	£8·50		

The prices listed in the table are reduced by a further 10% for cash. Find:

a the extra reduction on each item

b the reduced price for each item.

Round up to the nearest pence.

4 The cash price of a TV set is £320·00. If paid for over 12 months there is a deposit of £50 and a hire-purchase charge of 8%.

    a How much HP is paid?

    b What is the total cost buying on HP?

    c How much would each monthly payment be?

5 How much discount would you receive if goods priced £57 had a 40% discount?

6 Footballs were marked 15% off. If a boy paid £4·25 for a football, what was the usual price?

7 Bank interest is 6%. How much interest would you receive on:

    a £174·50   **b** £2,456·50   **c** £24,496·00

8 76% of a group of 50 children passed a cycling test. How many children failed?

9 What should be paid for a £2,475 car on sale at $33\frac{1}{3}$% discount?

10 3471 is 75% of which number?

Ratio

A **1** Express as ratios.

a $\frac{11}{20}$ $\frac{9}{10}$ $\frac{4}{5}$ $\frac{12}{15}$ $\frac{21}{30}$ $\frac{7}{8}$ $\frac{11}{12}$ $\frac{19}{25}$ $\frac{33}{100}$ $\frac{29}{50}$

b 40%   35%   27% $12\frac{1}{2}$%   75%   30%   16%   14%

2 **a** Change to vulgar fractions.

3:4   6:7   27:40   33:50   23:100   7:8

b Change to percentages.

7:20   9:50   13:25   18:25   3:10   23:100

B **1** a _____

b _____

c _____

d _____

Give these ratios.

a:b   a:c   a:d   b:c   b:d   c:d

d:a   d:c   d:b   b:a   c:a   c:b

2 Measure and draw these lines. Divide each up to the given ratio.

a _____   a 4:3

b _____   b 5:6

c _____   c 1:3

d _____   d 4:1

C **1** Give the gradients for each of the following by measuring the unknown sides.

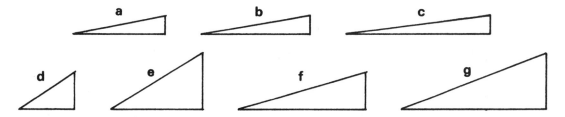

2 Draw slopes to show these gradients.

1:7   1:9   1:4   1:3   1:10

2:5   3:4   3:5   3:10   7:10

Ratio and proportion

A Work out the rise if the gradient is:

1:5   distance travelled 400 m   1:20   distance travelled 850 km

1:9   distance travelled 720 m   1:10   distance travelled 215 m

2:3   distance travelled 300 km   2:25   distance travelled 925 m

3:8   distance travelled 1 km   4:15   distance travelled 105 km

B **1** Divide each quantity into two parts in the ratio given.

    a 600 g ratio 1:5   **b** 45 children ratio 4:5

    c 24 sweets ratio 1:3   **d** 750 kg ratio 2:3

    e 110 trees ratio 10:1   **f** 60 oranges ratio 5:7

   2 Each quantity has been broken into two parts. Give the ratio.

72	48 and 24	91 m	39 m and 52 m
144p	84p and 60p	£2·75	£1·50 and £1·25
150 g	50 g and 100 g	2200	800 and 1400
400 l	250 l and 150 l	240 cm	150 cm and 90 cm

C **1** A car rises 20 metres on a road with a gradient of 1:7. How far has it travelled?

   2 On a length of railway track of 3 km the gradient is 1:30. How many metres does the track rise?

   3 A tractor rises $2\frac{1}{2}$ m travelling up a 10 m ramp. What is the gradient?

D **1** £25 is shared between William and Ari so that Ari has four times as much as William. How much do they each have?

   2 If 14 days holiday costs £350·00, how much would it cost for:

    a 10 days?   **b** 15 days?   **c** 24 days?

Capacity

A Complete:

1
2471 ml =   l   3016 ml =   l   743 ml =   l
76 ml =   l   126 ml =   l   2354 ml =   l

2
2·436 l =   ml   0·125 l   =   ml   0·637 l =   ml
0·047 l =   ml   2·071 l =   ml   0·007 l =   ml

3 Write as fractions:

500 ml =   l   400 ml =   l   450 ml =   l
750 ml =   l   850 ml =   l   350 ml =   l
4270 ml =   l   3680 ml =   l   1550 ml =   l
6150 ml =   l   2250 ml =   l   10 950 ml =   l

4
$4\frac{2}{5}$ l =   ml $6\frac{1}{4}$ l =   ml $\frac{7}{10}$ l =   ml

$\frac{7}{20}$ l =   ml $1\frac{4}{5}$ l =   ml $\frac{9}{25}$ l =   ml

$5\frac{3}{4}$ l =   ml $9\frac{239}{250}$ l =   ml $\frac{113}{250}$ l =   ml

$2\frac{3}{10}$ l =   ml $\frac{9}{10}$ l =   ml $12\frac{3}{5}$ l =   ml

5 Use the signs > or < to complete:

0·38 l   38 ml   4·06 l   4600 ml $1\frac{3}{5}$ l   1500 ml
$3\frac{3}{10}$ l   3030 ml $9\frac{7}{20}$ l   9035 ml   0·035 l   350 ml

B

1
2·36 l + 575 ml + $\frac{1}{2}$ l =   l   4·5 l + $6\frac{1}{4}$ l + 2 l 76 ml =   l

5·06 l + 0·027 l + $\frac{3}{4}$ l =   l   19·5 l + $6\frac{7}{10}$ l + 5420 ml =   l

2
6·257 l – $5\frac{3}{10}$ l =   ml $12\frac{1}{2}$ l – 7396 ml =   l

8 l 27 ml – 7 l 54 ml =   ml   21 l 16 ml – 9·239 l =   l

3
47·213 l × 6 =   l   3 l 60 ml × 5 =   l

41 271 ml × 9 =   l $7\frac{9}{25}$ l × 8 =   l

4
12·864 l ÷ 4 =   l $51\frac{3}{4}$ l ÷ 9 =   l
3 l 475 ml ÷ 5 =   l   515·252 l ÷ 12 =   l

Mass

A **1** Write as decimal fractions.

$2\frac{3}{4}$ kg $1\frac{4}{5}$ kg $\frac{7}{10}$ kg $\frac{9}{20}$ kg $3\frac{13}{25}$ kg $6\frac{27}{1000}$ kg

$3\frac{3}{100}$ kg $8\frac{1}{4}$ kg $\frac{1}{2}$ kg $\frac{11}{20}$ kg $1\frac{3}{5}$ kg $6\frac{13}{1000}$ kg

2 Write as vulgar fractions.

4·750 kg   0·850 kg   0·950 kg   3·4 kg   4·7 kg

0·925 kg   0·050 kg   4·45 kg   9·3 kg   0·25 kg

3 Which of these weights are greater than $\frac{1}{2}$ kg?

475 g   620 g   510 g   0·375 kg   0·601 kg   0·356 kg

4 Which of these weights are less than $\frac{3}{4}$ tonne?

675 kg   1046 kg   0·6 t   0·76 t   432 kg   0·78 t

5 Write these weights in order of size – smallest first.

1 t   1200 kg   2500 g   750 kg   0·75 kg   2·524 kg

6 Write these weights in tonnes and kilograms.

$3\frac{3}{4}$ t   6·9 t   11·75 t   6·85 t $8\frac{7}{10}$ t $5\frac{4}{5}$ t

7 Write these weights in tonnes.

6 t 27 kg   11 t 9 kg   4 t 263 kg   347 kg   62 kg

B Give your answers to the following as decimal fractions of tonnes or kilograms.

1   4·25 t + $2\frac{1}{2}$ t + 3472 kg $3\frac{1}{2}$ t + 704 kg + 925 kg

      8 kg 40 g + 800 g + 6 kg 50 g   1900 g + 700 g + 36·56 kg

2   27·35 kg − $25\frac{3}{4}$ kg   9·340 kg − 8 kg 709 g

      342 t − 276 t 78 kg $4\frac{1}{2}$ kg − 3124 g

3   7 kg 164 g × 7 $14\frac{1}{2}$ t × 12   6 kg 507 g × 8   35 t 176 kg × 10

4   35·208 t ÷ 9 $72\frac{3}{4}$ kg ÷ 6   111 t 5 kg ÷ 5   30 kg 800 g ÷ 7

Length

A **1** Write as kilometres.

3 km 27 m   21 km 345 m   432 m   3400 cm   12 000 mm

5300 cm   6 km 5 m   2475 m   26 m   12 km 35 m

2 Write as metres.

6 m 26 cm   5 m 6 cm   21 m 27 mm   45 mm   96 mm

12 m 175 mm   16 m 7 cm   2141 mm   21 m 6 mm   7 mm

3 Write as decimals.

$6\frac{1}{4}$ km $11\frac{1}{2}$ km $17\frac{3}{4}$ km $3\frac{7}{10}$ km $2\frac{4}{5}$ km $\frac{11}{20}$ km

$8\frac{3}{5}$ m $12\frac{3}{10}$ m $2\frac{7}{20}$ m $12\frac{3}{4}$ m $11\frac{9}{20}$ m $14\frac{9}{10}$ m

4 Write as vulgar fractions.

12·5 km   26·75 km   32·7 km   45·05 km   16·015 km

9·6 m   13·45 m   0·25 m   18·55 m   0·95 m

5 Write as kilometres and metres.

14·75 km   6·234 km   8·632 km   12·275 km   12·7 km

6·45 km   10·035 km $11\frac{4}{5}$ km $8\frac{9}{10}$ km   9·6 km

B **1**   2 km 40 m + 2·725 km + 3·2 km =   km

$2\frac{3}{4}$ km + 3·55 km + 732 m =   km

5 cm 8 mm + 10 cm 9 mm + 17 cm =   cm   mm

3·6 cm + 9 mm + 6·9 cm =   cm

2 73 cm − 635 mm =   cm   106 km − 7·432 km =   km

12·9 m − 16·2 cm =   m $115\frac{3}{4}$ km − 945 m =   km

3   3 m 75 cm × 4 =   m   4·125 cm × 6 =   cm

127 cm 7 mm × 11 =   cm   mm   73 m 9 mm × 9 =   m   mm

14·325 km × 7 =   km   0·375 m × 5 =   m

4 196 cm 5 mm ÷ 5 =   cm $325\frac{1}{2}$ m ÷ 7 =   m

2·376 km ÷ 6 =   km   698·4 cm ÷ 8 =   cm

41 m 4 cm ÷ 9 =   m   45 m 4 mm ÷ 4 =   m

Measures – problems

1   A box full of packets weighs 33 kg. If the box weighs $1\frac{1}{2}$ kg, how many 350 g packets does it hold?

2   Find the total length of four remnants of material measuring 9·6 cm, 19·27 cm, 26·51 cm and 32·75 cm.

3   How much orange juice is needed to fill 24 bottles each holding 725 ml?

4   A storage tank holds 10 000 litres of petrol. 3670 l is sold on Monday and 4165 l on Tuesday. How much is left in the tank?

5   A car is bought having been driven 11 345·7 km. After 1 year it has gone 19 462·9 km. What distance was covered in the first year?

6   One loaded lorry weighs 4 t 29 kg. Another weighs 3 t 125 kg. What is the difference in weight?

7   A school kitchen uses 120 kg of potatoes each day. How many kilograms are used in four weeks?

8   How many 250 ml cartons can be filled from 2400 l?

9   36 poles are fixed 1·75 m apart. How much wire is required to link them?

10   The daily deliveries of milk to a school are 290 l, 280 l, 325 l, 350 l, 375 l. How much is delivered in the week?

11   Four salesmen travel the following distances in a year: 12 475 km, 16 954 km, 18 627 km and 20 432 km. What was the average distance travelled?

12   A lorry carried 13 t 75 kg on the first day, 14 t 17 kg on the second day and 16 t 96 kg on the third day. What was the total weight for the three days?

Area and perimeter

A Find: **a** the area in mm²   **b** the perimeter in mm

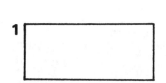  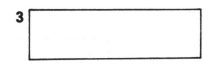

B Find the perimeter of rectangles which measure:

   a   20·4 cm by 7 cm   **b**   36·4 mm by 15 mm   **c**   9 m × 12·6 m

   d   12·7 cm by 8·3 cm   **e**   914 mm by 131 mm   **f**   11·6 m × 15·3 m

C Complete the table of measurements of rectangles.

length	15 cm	16 cm			25 cm	25 cm	
width	19 cm		12 cm	18 cm		34 cm	23 cm
perimeter			62 cm		80 cm		
area		176 cm²		306 cm²			598 cm²

D Find: **a** the perimeter   **b** the area.

  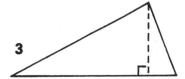

E Find the area of triangles which measure:

base	height		base	height
17 cm	15 cm		23·5 cm	12 cm
22 cm	20 cm		9·5 m	7 m
75 mm	27 mm		10·75 cm	14 cm

F Complete this table of measurements of triangles.

base	12 cm	15 cm	12 cm		24 cm	30·5 cm	
height		16 cm		15 cm	9 cm		32 cm
area	60 cm²		72 cm²	112·5 cm²		122 cm²	160 cm²

Area and perimeter

Find the area of the unshaded parts.

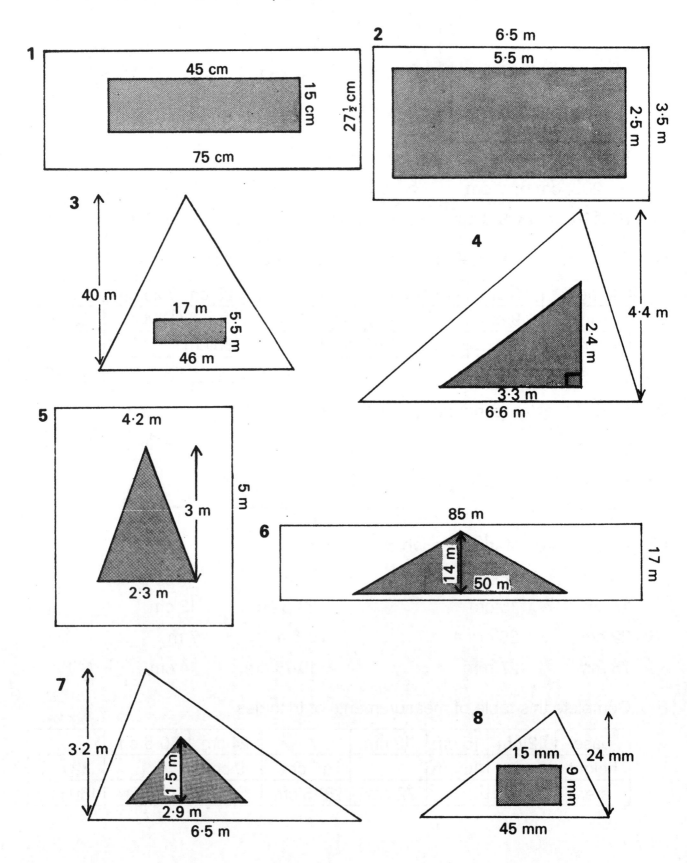

1
45 cm
15 cm
27½ cm
75 cm

2
6·5 m
5·5 m
2·5 m
3·5 m

3
40 m
17 m
5·5 m
46 m

4
4·4 m
2·4 m
3·3 m
6·6 m

5
4·2 m
3 m
5 m
2·3 m

6
85 m
14 m
50 m
17 m

7
3·2 m
1·5 m
2·9 m
6·5 m

8
15 mm
24 mm
9 mm
45 mm

Area

Find the total area of each of these shapes.

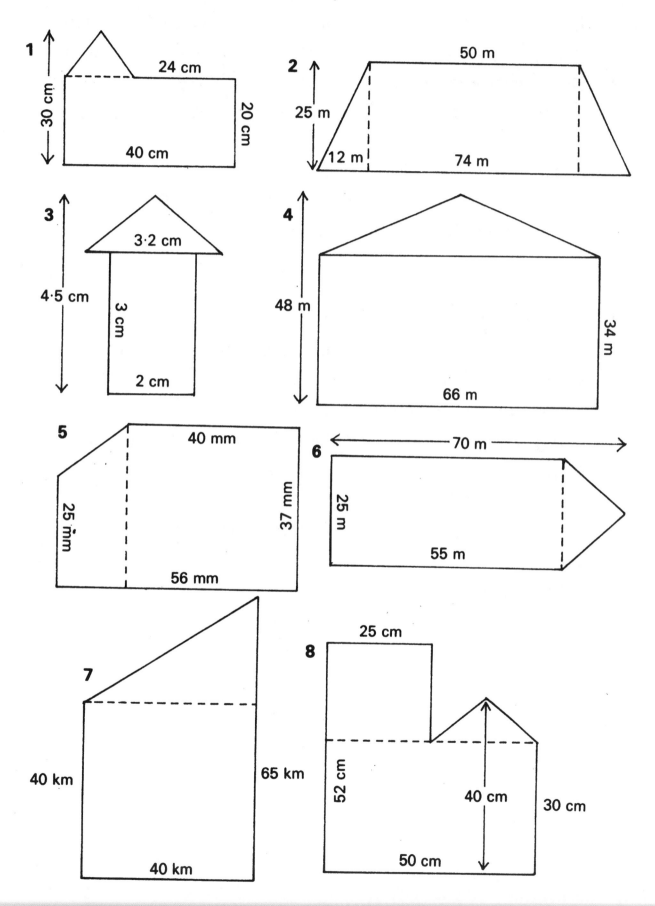

1

30 cm
24 cm
20 cm
40 cm

2

50 m
25 m
12 m
74 m

3

3·2 cm
4·5 cm
3 cm
2 cm

4

48 m
66 m
34 m

5

40 mm
25 mm
37 mm
56 mm

6

70 m
25 m
55 m

7

40 km
65 km
40 km

8

25 cm
52 cm
40 cm
30 cm
50 cm

45

Circles – circumference

A **1** Draw circles with a radius of:

2 cm $\quad\quad$ $5\frac{1}{2}$ cm $\quad\quad$ 3 cm $\quad\quad$ $4\frac{1}{2}$ cm $\quad\quad$ $1\frac{1}{2}$ cm

2·4 cm $\quad\quad$ 3·2 cm $\quad\quad$ 2·9 cm $\quad\quad$ 4·4 cm $\quad\quad$ 1·8 cm

2 Draw circles with a diameter of:

5·2 cm $\quad\quad$ 10 cm $\quad\quad$ $7\frac{1}{2}$ cm $\quad\quad$ 8·5 cm $\quad\quad$ 11 cm $\quad\quad$ 6·8 cm

B Give the radius when the diameter is:

1 11·2 m $\quad\quad$ 9·6 cm $\quad\quad$ 12·7 m $\quad\quad$ 15·2 m $\quad\quad$ 16·9 cm $\quad\quad$ 13·5 cm

2 Give the diameter when the radius is:

$7\frac{3}{4}$ cm $\quad\quad$ $15\frac{1}{2}$ cm $\quad\quad$ $2\frac{1}{4}$ m $\quad\quad$ $6\frac{7}{10}$ m $\quad\quad$ $4\frac{3}{5}$ cm $\quad\quad$ $12\frac{1}{2}$ m

C **1** **Circumference of a circle = 2πr**

Using $3\frac{1}{7}$ as the value of π, find the circumference if the radius is:

$2\frac{3}{4}$ cm $\quad\quad$ $6\frac{3}{4}$ cm $\quad\quad$ $4\frac{1}{2}$ cm $\quad\quad$ 9 cm $\quad\quad$ $12\frac{1}{2}$ cm $\quad\quad$ $5\frac{1}{2}$ cm

2 **Circumference of a circle = πd**

Using $3\frac{1}{7}$ as the value of π, find the circumference if the diameter is:

7 cm $\quad\quad$ $6\frac{1}{2}$ cm $\quad\quad$ 15 cm $\quad\quad$ 19 cm $\quad\quad$ $7\frac{3}{4}$ cm $\quad\quad$ $14\frac{1}{2}$ cm

D **1** Using π as 3·14, find the circumference when the radius is:

2·7 m $\quad\quad$ 3·4 cm $\quad\quad$ 6·9 m $\quad\quad$ 18·5 cm $\quad\quad$ 16·7 cm

2 Using π as 3·14, find the circumference when the diameter is:

29·8 m $\quad\quad$ 47·2 cm $\quad\quad$ 35·6 cm $\quad\quad$ 18·5 m $\quad\quad$ 0·75 m

Circles – area and circumference

A **Area of a circle = πr^2.** Use $3\frac{1}{7}$ as the value of π.

    1 Find the area of the circle if the radius is:

        7 cm   14 cm   28 cm   35 cm $5\frac{1}{4}$ cm $7\frac{7}{8}$ cm

    2 Find the area of a circle if the diameter is:

        $3\frac{1}{2}$ cm $10\frac{1}{2}$ cm   21 cm $24\frac{1}{2}$ cm $31\frac{1}{2}$ cm   35 cm

B Use 3·14 as the value of π.

    1 Find the area of the circle when the radius is:

        3 cm   9 cm   6 cm   11 cm   4 cm   7 cm

    2 Find the area of the circle if the diameter is:

        28 cm   24 cm   36 cm

        22 cm   48 cm   10 cm

C **1** Complete the table using 3·14 as the value of π.

diameter (in cm)	12·8					1·6	
radius (in cm)		5·1			1·4		
circumference (in cm)			28·26				50·24
area (in cm²)				78·5			

    2 Find the area and perimeter of these shapes.

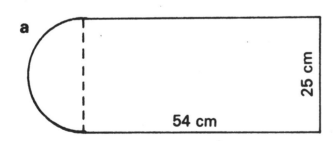

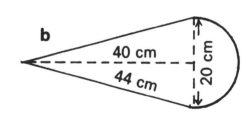

Volume of cubes and cuboids

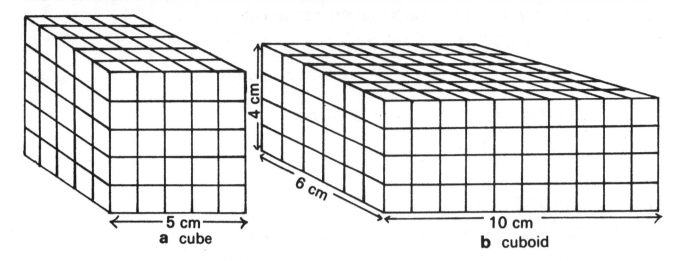

a cube   b cuboid

A   **1**   What is the volume of cube **a** in cm³?

      2   Complete these tables of volumes of cubes, using the formula
          v = l × b × h.

length of side	4 cm	6 cm	11 cm	9 cm	10 cm	$4\frac{1}{2}$ cm
volume						

length of side	$6\frac{1}{2}$ cm	$10\frac{1}{2}$ cm	$11\frac{1}{2}$ cm	$12\frac{1}{2}$ cm	$8\frac{1}{2}$ cm
volume					

B   **1**   What is the volume of the cuboid **b** above in cm³?

      2   Complete this table of volumes of cuboids, using the formula
          v = l × b × h.

length	7 cm	$10\frac{1}{2}$ cm	$15\frac{1}{2}$ cm	8 cm	22 cm	$12\frac{1}{2}$cm	15 cm
breadth	$4\frac{1}{2}$ cm	6 cm	$3\frac{1}{2}$ cm	$4\frac{1}{2}$ cm	$3\frac{1}{2}$ cm	10 cm	8 cm
height	4 cm	2 cm	4 cm	$2\frac{1}{2}$ cm	$2\frac{1}{4}$ cm	$2\frac{1}{2}$ cm	$3\frac{1}{2}$ cm
volume							

      3   Write the formula used to find:

          a   length when the breadth, height and volume are known.

          b   breadth when the length, height and volume are known.

          c   height when the breadth, length and volume are known.

Volume – triangular prisms and cylinders

A **1**   What is the volume in cm³ of prism **x** and prism **y**?

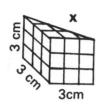 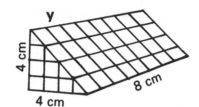

2   Find: **a** the area of triangle ABC in **r**
         b the volume of prism **r**.

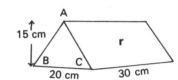 

3   Find the volume of prism **s**, above.

4   Find the volume of the triangular prisms in the table. Remember:

$$v = b \times \tfrac{1}{2} \text{ perpendicular ht} \times l$$

triangular prism	a	b	c	d	e	f	g	h	i	J
base (b) in cm	6	9	10	15	23	25	35	50	28	30
perpendicular height (ht) in cm	5	$4\tfrac{1}{2}$	7	$9\tfrac{1}{2}$	12	10	16	15	20	11
length (l) in cm	15	10	20	15	25	16	12	30	15	25
volume (v) in cm³										

B

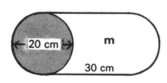 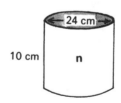

1   Find: **a** the area of the shaded part of **m**,

         b the volume of the cylinder **m**.   Use $\tfrac{22}{7}$ as the value of π.

2   Find the volume of cylinder **n** in the same way.

3   Use $\pi r^2 l$ to find the volume of these cylinders. Use 3·14 as the value of π.

cylinder	a	b	c	d	e	f
radius (r) in cm	4	7	3	9	2	6
length (l) in cm	10	20	5	12	8	30
volume (v) in cm³						

Volume – assorted shapes

Find the volume of these shapes.

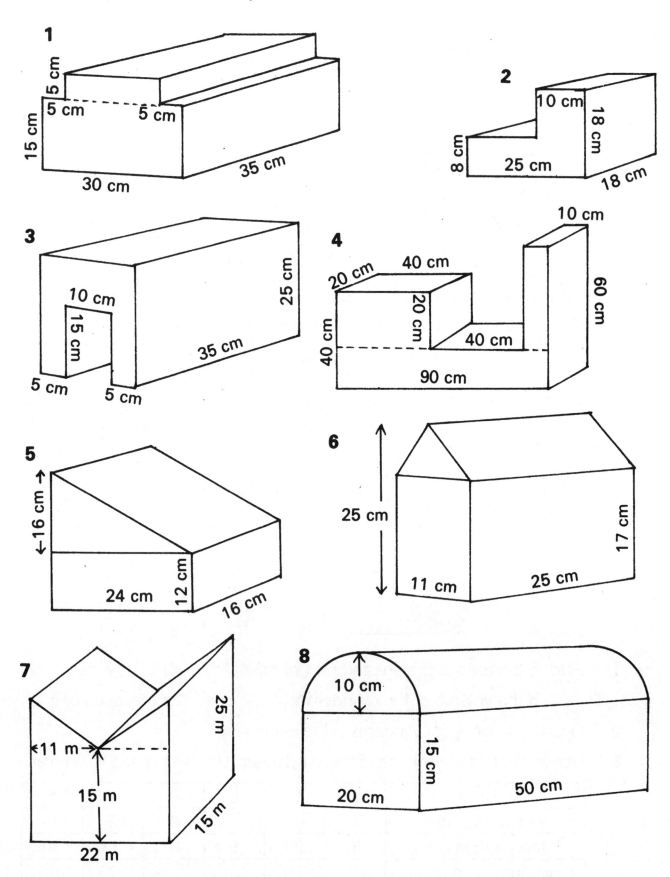

1
5 cm
5 cm
5 cm
5 cm
15 cm
30 cm
35 cm

2
10 cm
18 cm
8 cm
25 cm
18 cm

3
10 cm
15 cm
5 cm
5 cm
25 cm
35 cm

4
10 cm
20 cm
40 cm
20 cm
40 cm
40 cm
60 cm
90 cm

5
16 cm
24 cm
12 cm
16 cm

6
25 cm
11 cm
25 cm
17 cm

7
11 m
15 m
15 m
25 m
22 m

8
10 cm
15 cm
20 cm
50 cm

Scale drawing

A **1** A map has a scale of 1 mm to 10 km.

    a Draw lines using this scale to represent these distances from London.

        Manchester 290 km   Liverpool 300 km   Southampton 250 km

              Edinburgh 600 km   Aberdeen 780 km

    b Using the same scale, what is the real distance of these places from London?

        Penzance 4·5 cm   Glasgow 6·2 cm   Cardiff 2·5 cm

        York 3·1 cm   Leeds 3 cm

B This wall unit is drawn to a scale of 1 cm to 20 cm. Measure the lengths indicated and give their real lengths.

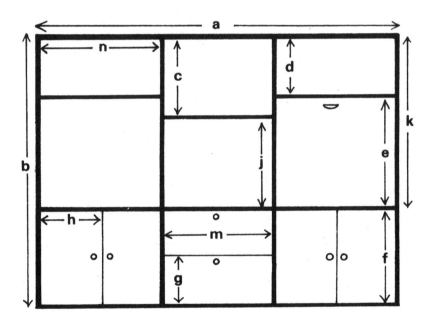

C Give the scale used for each of these lines.

    ___60 km___

    _____14 km_____

    _____250 km_____

    _____320 km_____

    _____300 km_____

Scale drawing

A   This is a sketch of an allotment. Draw it accurately to a scale of 0·2 cm to 1 m.

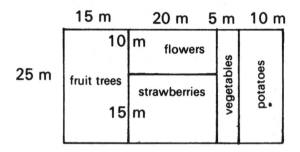

B   Find out how far each ship is from the lighthouse.

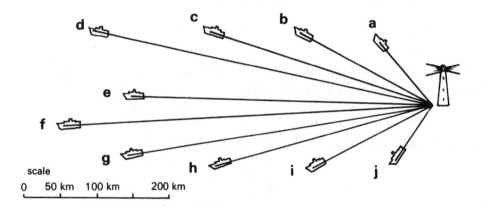

C

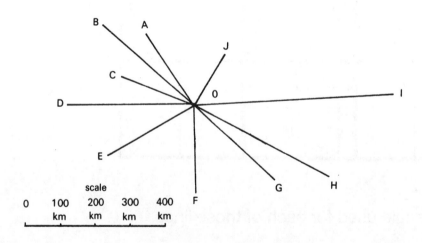

1   A, B, C, D, E, F, G, H, I, J are airports. How far is each airport from 0?

2   How far would each distance be if the scale of the map was:

    a   1 cm to 50 km?   **b**   1 cm to 20 km?   **c**   1 cm to 40 km?

    d   1 cm to 200 km?

Angles

A **1** Use a protractor to draw these angles.

| 62° | 27° | 54° | 83° | 35° | 75° | 45° | 15° |

| 95° | 125° | 160° | 135° | 170° | 140° | 115° | 110° |

2 Measure and record the two angles on each line.

a 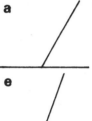 b 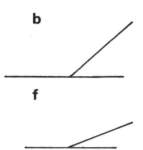 c 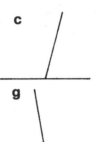 d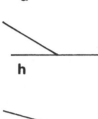

e f g h

3 On separate straight lines draw angles of:

a 35° **b** 65° **c** 130° **d** 145° **e** 25° **f** 55°

Measure and record the size of the supplementary angles you have drawn.

4 Calculate the size of the unmarked angle.

a b c d

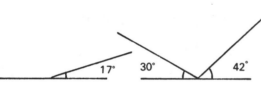 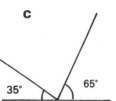

17° 30° 42° 35° 65° 85°

B **1** Measure and record the size of each angle.

a b c

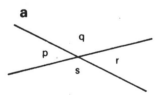 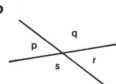 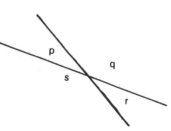

2 Without using a protractor, record the size of the angles:

a x, y, z **b** a, b, c **c** r, s, t.

a b c

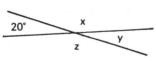 

20° x y z 32° a b c 56° r t s

Angles

A 1 Measure and record the size of the angles in each of these triangles.

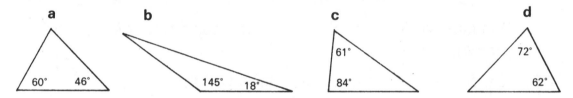

2 a On a base line AB draw triangles with the following base angles.

30° 40°	75° 50°	25° 80°	90° 45°	65° 30°
105° 45°	125° 20°	120° 35°	55° 60°	70° 60°

  b Measure and record the unknown angle in each triangle.

3 Without using a protractor, calculate the size of the missing angle in these triangles.

a 60° 46° **b** 145° 18° **c** 61° 84° **d** 72° 62°

B 1 Measure and record the size of the angles in each of these quadrilaterals.

a P Q S R **b** P Q S R **c** Q P S R

2 Draw quadrilaterals using the following sets of three angles, then measure and record the unknown angle.

75° 110° 95°	45° 120° 70°	50° 60° 115°
30° 130° 150°	55° 75° 125°	80° 95° 100°

3 Calculate the size of the missing angle in these quadrilaterals.

  a 105° 95° 60°   **b** 150° 145° 45°   **c** 200° 70° 60°

Bearings

A   Plot the following bearings.

230°   140°   85°   310°   100°   210°   75°   330°

B   These ships' courses are drawn to a scale of 1 cm to 200 km. Give the bearing and distance for each part of the voyage, then give the bearing and distance to reach home.

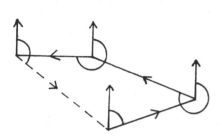
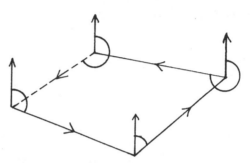

C   Plot these courses and then give the bearing and distance back to base (scale 1 cm to 100 km).

1	bearing	distance	**2**	bearing	distance
	70°	350 km		265°	300 km
	350°	250 km		20°	300 km
	270°	550 km		110°	550 km

D

☐• cinema   ☐• school

•Peter

•Freya

| scale 5 cm to 1 km |

        •Alana   •Jack

•John   •Olivia

Give the approximate bearing each child must take to reach:

a   school   **b**   the cinema.

How far does each child live from:

a   school   **b**   the cinema?

Timetables

A **1** Buses leave Manchester for Liverpool every forty minutes. Give the departure times from Manchester from 09:00 h to 15:00 h.

      2 The journey takes 1 h 15 min. Give the arrival times at Liverpool.

      3 The bus crew have a quarter of an hour rest before starting the return journey. Give the departure time for each bus from Liverpool.

      4 Give the arrival times at Manchester.

B **1** Boat trips on the river at Chester commence at 09:15. Boats leave at thirty minute intervals for a forty minute sail. Complete this timetable for one of the boats.

depart	09:15	10:15						
arrive	09:55							

      2 How many trips does the boat make each day?

      3 What time does the last boat sail?

      4 A school party has to leave Chester at 15:30. What is the latest time they can take a boat trip?

      5 The boat holds 72 people. If it is full on each trip how many people will be carried in one day?

      6 On a certain day, one third of the passengers carried were children. If the fare is 40p per adult and 20p per child how much money was taken on this day?

C A miniature railway runs twenty minute round trips commencing at 09:00 until 12:15. If there is a 5 minute wait at the end of each trip, draw up the railway timetable for the train.

Speed

A   A car travels at an average speed of 60 km/h.

    1   How far will it travel in:

        a   5 h?   **b**   6·5 h?   **c**   8·25 h?   **d**   2·75 h?

        e   4·3 h?   **f**   2 h 12 min?   **g**   4 h 35 min?   **h**   1 h 50 min?

    2   How long will it take to travel:

        a   250 km? **b**   320 km? **c**   262 km? **d**   455 km? **e**   610 km?

B   A coach travels 4800 km.

    1   How many hours was the coach driven at an average speed of:

        a   20 km/h?   **b**   50 km/h?   **c**   100 km/h?   **d**   30 km/h?

        e   80 km/h?   **f**   150 km/h?   **g**   60 km/h?   **h**   120 km/h?

    2   At what speed did the coach travel if it was driven for:

        a   50 h?   **b**   75 h?   **c**   60 h?   **d**   40 h?   **e**   80 h?

C   **1**   Convert to km/h.

        a   5·5 km in 12 min   **b**   4·25 km in 5 min   **c**   35 km in 2 min

        d   12·25 km in 4 min   **e**   10·5 km in $1\frac{1}{2}$ min   **f**   20·5 km in 20 min

    2   Convert to m/s.

        a   0·3 km in $\frac{1}{2}$ min   **b**   0·6 km in $\frac{1}{4}$ min   **c**   0·1 km in $\frac{1}{6}$ min

        d   0·4 km in $\frac{2}{3}$ min   **e**   0·9 km in $\frac{3}{4}$ min   **f**   0·3 km in $\frac{5}{6}$ min

    3   Convert to km/h.

        a   100 m/s   **b**   230 m/s   **c**   192 m/s

        d   150 m/s   **e**   210 m/s   **f**   80 m/s

    4   Convert to m/s.

        a   180 km/h   **b**   54 km/h   **c**   126 km/h

        d   162 km/h   **e**   27 km/h   **f**   252 km/h

Speed

A

flight	distance	time	av. speed
a	3500 km	$3\frac{1}{2}$ h	
b	2400 km	$2\frac{1}{2}$ h	
c		3 h 20 min	1200 km/h
d	2240 km	2 h 40 min	
e	3680 km		960 km/h
f	4400 km	$5\frac{1}{2}$ h	

1 Copy and complete this table of air flights.

2 Which two planes travel at the same speed?

3 List the planes in order of speed – fastest first.

B The speed of very fast aircraft is sometimes measured in Mach numbers. Mach 1 is 1200 km/h.

1 Convert these Mach numbers to km/h.

Mach 2   Mach 3   Mach 4   Mach 5   Mach 6

Mach 2·5   Mach 4·75   Mach 3·2   Mach 4·1   Mach 6·3

2 Convert these km/h to Mach numbers.

7440 km/h   2880 km/h   1980 km/h   5700 km/h

6300 km/h   3960 km/h   2850 km/h   1320 km/h

Graphs

A

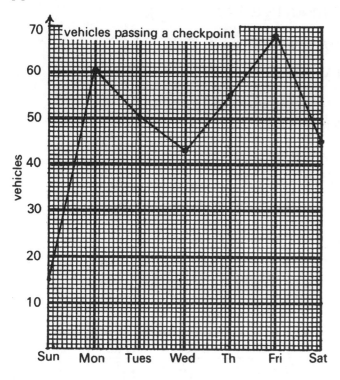

1 Which day has:

    a most vehicles?   **b** least vehicles?

2 What is the difference between the highest and lowest totals?

3 What was the average number of vehicles per day?

4 On which day was the total 25% of Monday's total?

5 On which day was the total:

    a $\frac{3}{4}$ **b** $\frac{5}{6}$

    of Monday's total?

6 Which day's total was 3 times that of Sunday?

7 How many totals were more than the average?

8 How many totals were less than average?

B Draw the graph which shows the number of fine days each month in one year.

Jan	Feb	Mar	Apr	May	June	July	Aug	Sept	Oct	Nov	Dec
7	15	17	18	25	26	20	19	22	20	13	9

C

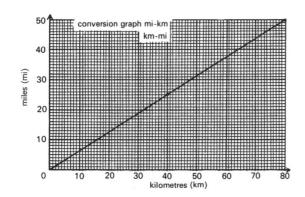

Change to kilometres			
5 mi	15 mi	30 mi	40 mi
$17\frac{1}{2}$ mi	$27\frac{1}{2}$ mi	$42\frac{1}{2}$ mi	$22\frac{1}{2}$ mi
Change to miles			
24 km	48 km	32 km	56 km
20 km	52 km	12 km	60 km

D Draw conversion graphs using this information.

kilograms	1	4	7	old pence	2.4	24	36
pounds	2.2	8.8	15.4	new pence	1	10	15

Comparison graphs

A

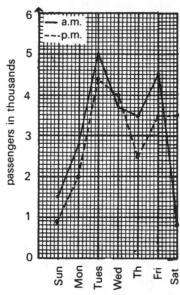

The graph shows the number of people using a bus station mornings and evenings.

1 How many travelled each morning?

2 How many travelled each evening?

3 How many travelled on the 7 mornings altogether?

4 How many travelled on the 7 evenings altogether?

5 What is the difference in these totals?

6 On which mornings did the total exceed $3\frac{1}{2}$ thousand?

B   Draw a graph to record this information.

midday	8°C	4°C	9°C	6°C	4°C	0°C	7°C
midnight	3°C	–4°C	2°C	–1°C	–2°C	–5°C	3°C
day	Sun	Mon	Tue	Wed	Thurs	Fri	Sat

C

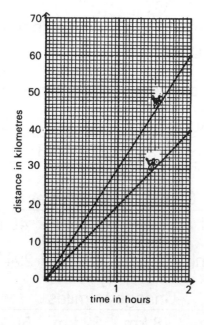

The graph shows a comparison between the distances travelled by a cyclist and scooter driver.

1   What is the speed in km/h of:

    a   the cyclist?   **b**   the scooter?

2   How far has each travelled after:

    a $\frac{1}{4}$ h?   **b** $\frac{1}{2}$ h?

    c $1\frac{1}{4}$ h?   **d** $1\frac{1}{2}$ h?

    e $1\frac{3}{4}$ h?

3   How much farther has the scooter travelled after:

    a   12 min?   **b**   21 min?

    c   39 min?   **d**   1 h 9 min?

    e   1 h 21 min?   **f**   1 h 57 min?

4   If the scooter breaks down after one hour, how long would it take the cyclist to catch up?

D   Draw a graph to compare these speeds – 80 km/h and 60 km/h.

Co-ordinates

A **1**

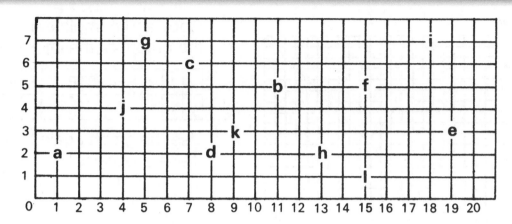

Write the co-ordinates of the points **a** to **l**.

2 On a piece of squared paper plot the following co-ordinates listing the points **a** to **l**.

a (0, 1) **b** (2, 7) **c** (3, 6) **d** (4, 4) **e** (6, 7) **f** (9, 5)

g (2, 8) **h** (5, 3) **i** (4, 5) **j** (9, 2) **k** (7, 4) **l** (8, 3)

B

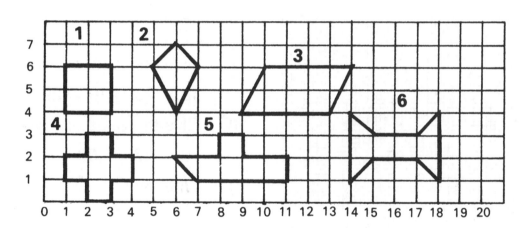

1 List the co-ordinates needed to plot each shape.

2 On a piece of squared paper plot these co-ordinates. They will give five letters to make a word.

a (2, 1) (2, 5) (1, 5) (1, 6) (4, 6) (4, 5) (3, 5) (3, 1) (2, 1)

b (5, 1) (5, 6) (6, 6) (6, 4) (7, 4) (7, 6) (8, 6) (8, 1) (7, 1) (7, 3) (6, 3) (6, 1) (5, 1)

c (9, 1) (9, 6) (10, 6) (10, 1) (9, 1)

d (11, 1) (11, 6) (12, 6) (13, 4) (13, 6) (14, 6) (14, 1) (13, 1) (12, 3) (12, 1) (11, 1)

e (15, 1) (15, 6) (16, 6) (16, 4) (17, 6) (18, 6) (17, 4) (17, 3) (18, 1) (17, 1) (16, 3) (16, 1) (15, 1)

Co-ordinates

A

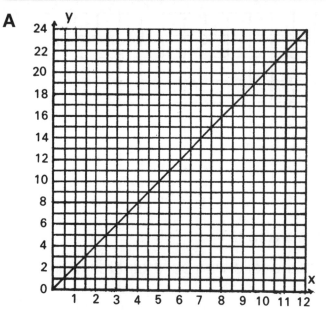

1 Copy and complete the table. Use the graph for the information.

x	1	2	3	4	5	6	7	8	9	10	11	12
y												

2 Use the graph again to answer these questions.

$1\frac{1}{2} \times 2$ $7\frac{1}{2} \times 2$ $10\frac{1}{2} \times 2$ $6\frac{1}{2} \times 2$

$9\frac{1}{2} \times 2$ $3\frac{1}{2} \times 2$ $5\frac{1}{2} \times 2$ $11\frac{1}{2} \times 2$

$15 \div 2$ $21 \div 2$ $23 \div 2$ $9 \div 2$

$5 \div 2$ $13 \div 2$ $17 \div 2$ $19 \div 2$

B **1** Copy and complete the table. Use the graph for the information.

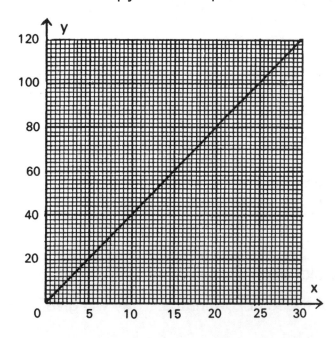

x	5		15	20		30
y		40			100	

2 Give the value of y when x is:

12   28   18   22   16   26

3 Give the value of x when y is:

70   36   108   86   50   28

4 What is:

9×4 21×4 29×4 16×4

13×4 28×4 19×4 11×4

$108 \div 4$ $92 \div 4$ $76 \div 4$

$116 \div 4$ $56 \div 4$ $104 \div 4$

C Complete these tables, then draw the graph to show the relationship between x and y.

x	1	2	5	10	15	20
y	8	16	40	80	120	160

x	1	2	5	10	15	20
y	7	14	35	70	105	140

Answers

Page 2 Notation

A **1a** 472 047 **b** four hundred and seventy-two thousand and forty-seven

2a 5 201 649 **b** five million, two hundred and one thousand, six hundred and forty-nine

3a 5 954 270 **b** five million, nine hundred and fifty-four thousand, two hundred and seventy

4a 6 640 897 **b** six million, six hundred and forty thousand, eight hundred and ninety-seven

5a 2 068 083 **b** two million, sixty-eight thousand and eighty-three

B 4 000 000, 7 000 000, 19 000 000, 2 000 000, 15 000 000;
500 000, 250 000, 750 000, 200 000, 100 000;
3 250 000, 2 750 000, 6 500 000, 2 600 000, 4 700 000;
100 000, 400 000, 700 000, 200 000, 600 000;
3 300 000, 2 500 000, 1 800 000, 3 400 000, 6 900 000

C Add 10 **a** 402 351 **b** 160 602 **c** 1 000 009
Add 100 **a** 634 901 **b** 503 016 **c** 1 271 314
Add 10 000 **a** 447 216 **b** 60 439 **c** 217 986
Add 100 000 **a** 1 021 741 **b** 943 216 **c** 2 041 276
Add 1000 **a** 1 500 211 **b** 1 377 416 **c** 2 000 271

D five thousand, two million,
nine thousand, fifty thousand;
ninety thousand, one million,
three tens, seven hundred thousand

E $\frac{1}{2}$ million **a** 499 999 **b** 499 990 **c** 499 900
$\frac{1}{4}$ million **a** 249 999 **b** 249 990 **c** 249 900
$\frac{3}{4}$ million **a** 749 999 **b** 749 990 **c** 749 900

300 000 **a** 299 999 **b** 299 990 **c** 299 900
350 000 **a** 349 999 **b** 349 990 **c** 349 900
4·2 million **a** 4 199 999 **b** 4 199 990 **c** 4 199 900
2·1 million **a** 2 099 999 **b** 2 099 990 **c** 2 099 900
8·3 million **a** 8 299 999 **b** 8 299 990 **c** 8 299 900
470 000 **a** 469 999 **b** 469 990 **c** 469 900
610 000 **a** 609 999 **b** 609 990 **c** 609 900

Page 3 Approximations

A 54 729 **a** 54 730 **b** 54 700 **c** 55 000
13 269 **a** 13 270 **b** 13 300 **c** 13 000
202 435 **a** 202 440 **b** 202 400 **c** 202 000
60 987 **a** 60 990 **b** 61 000 **c** 61 000
120 696 **a** 120 700 **b** 120 700 **c** 121 000
444 909 **a** 444 910 **b** 444 900 **c** 445 000
38 451 **a** 38 450 **b** 38 500 **c** 38 000
899 823 **a** 899 820 **b** 899 800 **c** 900 000
54 163 **a** 54 160 **b** 54 200 **c** 54 000
914 429 **a** 914 430 **b** 914 400 **c** 914 000

B 14·352 **a** 14·35 **b** 14·4 **c** 14
16·073 **a** 16·07 **b** 16·1 **c** 16
7·451 **a** 7·45 **b** 7·5 **c** 7
242·715 **a** 242·72 **b** 242·7 **c** 243
6·439 **a** 6·44 **b** 6·4 **c** 6
11·992 **a** 11·99 **b** 12·0 **c** 12
13·475 **a** 13·48 **b** 13·5 **c** 13
6·549 **a** 6·55 **b** 6·5 **c** 7
8·727 **a** 8·73 **b** 8·7 **c** 9
18·296 **a** 18·30 **b** 18·3 **c** 18

C 430 000, 520 000, 1 420 000, 800 000, 30 000;
1 430 000, 350 000, 2 070 000, 610 000, 410 000

D 15 000 000, 27 000 000, 5 000 000, 13 000 000;
7 000 000, 33 000 000, 11 000 000, 6 000 000

E 500 000, 800 000, 500 000, 1 400 000, 400 000;
5 300 000, 800 000, 400 000, 700 000, 700 000

F 5 416 299 **a** 5 416 000 **b** 5 420 000 **c** 5 400 000 **d** 5 000 000
14 767 345 **a** 14 767 000 **b** 14 770 000 **c** 14 800 000 **d** 15 000 000
4 636 921 **a** 4 637 000 **b** 4 640 000 **c** 4 600 000 **d** 5 000 000
12 392 906 **a** 12 393 000 **b** 12 390 000 **c** 12 400 000 **d** 12 000 000
7 439 899 **a** 7 440 000 **b** 7 440 000 **c** 7 400 000 **d** 7 000 000
12 930 914 **a** 12 931 000 **b** 12 930 000 **c** 12 900 000 **d** 13 000 000

Page 4 Addition and subtraction

A 540 334, 817 506;   957 402, 1 689 078;
1475, 3 312 753;   255 246, 11 815;
7 500 000, 335 998

B 273 704, 70 210;   1 182 601, 250 000; 147 858, 744 605;   131 464, 150 000; 11 321, 2 050 000

C **1** 767 066 **2** 376 850 **3** 193 541 **4** 250 971 **5** 7259 **6** 1 523 603 **7** 486 093 **8** 7991 **9** 4 791 476 **10** 235 893 **11** 1 309 807 **12** 958 146

Page 5 Multiplication and division

A **1** 3480, 14 754, 35 376, 4630, 42 189 **2** 39 231, 36 364, 88 083

B 3750, 4260, 17 430, 205 490; 158 130, 48 580, 392 220, 1 287 680

C 896, 1817, 3672, 7144, 1003, 1827

D 15 232, 16 686, 24 543, 35 070, 32 093, 14 744

E 13 975, 47 400, 226 412, 677 446

F 72 r 4, 176 r 1, 78 r 4, 78 r 1, 126 r 1, 116

G 115 r 6, 352 r 1, 693 r 9, 785 r 7, 1773

H 1680 r 11, 8923 r 4, 16 885 r 1, 3171 r 4

I 6, 2 r 25, 3 r 14, 4 r 1, 3 r 8, 3, 3 r 1

J 6 r 18, 8 r 25, 9, 9 r 17, 9 r 15, 9 r 30

K 17 r 5, 25, 46 r 15, 31 r 1, 26 r 27, 20 r 1

L 75 r 12, 70 r 1, 24 r 45, 73 r 33, 75 r 7

M 201 r 19, 178 r 1, 244 r 20, 415 r 1, 271 r 14

N 166 r 5, 43 r 15, 92 r 22, 404 r 10

Page 6 Long multiplication

A **1** 19 **a** 190 **b** 570 **c** 950 **d** 1520
26 **a** 260 **b** 780 **c** 1300 **d** 2080
25 **a** 250 **b** 750 **c** 1250 **d** 2000
73 **a** 730 **b** 2190 **c** 3650 **d** 5840
51 **a** 510 **b** 1530 **c** 2550 **d** 4080
46 **a** 460 **b** 1380 **c** 2300 **d** 3680
83 **a** 830 **b** 2490 **c** 4150 **d** 6640
92 **a** 920 **b** 2760 **c** 4600 **d** 7360
54 **a** 540 **b** 1620 **c** 2700 **d** 4320
31 **a** 310 **b** 930 **c** 1550 **d** 2480
2 35 **a** 3500 **b** 14 000 **c** 24 500
42 **a** 4200 **b** 16 800 **c** 29 400
75 **a** 7500 **b** 30 000 **c** 52 500
87 **a** 8700 **b** 34 800 **c** 60 900
96 **a** 9600 **b** 38 400 **c** 67 200
58 **a** 5800 **b** 23 200 **c** 40 600
29 **a** 2900 **b** 11 600 **c** 20 300
18 **a** 1800 **b** 7200 **c** 12 600
67 **a** 6700 **b** 26 800 **c** 46 900
72 **a** 7200 **b** 28 800 **c** 50 400

B **1** 17 500, 26 900, 72 700, 58 600, 39 200, 82 600
2 238 700, 101 000, 339 500, 276 300, 216 400, 259 200

C 37 050, 153 510, 65 160, 86 640, 85 600, 120 770

D 118 230, 240 950, 803 520, 110 760, 365 160, 757 120

E 115 440, 190 350, 514 710, 243 380, 603 250, 679 910

F 136 955, 256 186, 454 662, 215 358, 793 548, 134 070

G 388 832, 241 092, 429 128, 721 308, 499 372, 94 218;
72 848, 338 424, 290 346, 446 665, 122 094, 510 094

Page 7 Long division

A 1827 r 11, 2451 r 13, 3142 r 2, 1508 r 9; 2423 r 15, 2194 r 10, 4636 r 3, 3122 r 10

B 1550 r 21, 2619 r 3, 1418 r 21, 1213 r 10; 2190 r 22, 3412 r 23, 2131 r 7, 2005 r 1

C 731 r 13, 1851 r 29, 723 r 19, 2854 r 23; 1139 r 8, 1333 r 14, 2358 r 19, 463 r 35

D 2054 r 18, 755 r 6, 1220 r 21, 1029 r 22; 2666 r 7, 788 r 20, 1471 r 16, 2028 r 2

E 6 r 121, 5 r 52, 5 r 104, 6 r 29

F 6 r 101, 9 r 67, 5 r 149, 8 r 130

G 11 r 44, 15 r 206, 31 r 10, 24 r 184

H 128 r 47, 623 r 20, 320 r 173; 258 r 30, 313 r 86, 355 r 191

I 273 r 4, 258 r 1, 606 r 15; 1204 r 21, 153 r 72, 240 r 21; 37 r 109, 239 r 9, 619 r 14

J 231 r 3, 201 r 5, 160 r 87; 185 r 6, 852 r 8, 432 r 3

Page 8 Decimals

A **1** 0·12, 0·3, 0·017, 0·003, 0·7, 0·09, 0·9, 0·071, 0·07, 0·009
2 0·75, 0·52, 0·14, 0·95, 0·5, 0·26, 0·35, 0·44, 0·85, 0·25
3 4·019, 6·12, 8·45, 1·7, 12·03, 10·063, 2·68, 5·15

B **1** 136, 75, 1204, 7, 4071, 5001, 16, 9
2 5134·67, 43·6715, 643·571, 3·67154, 3·17645
3 6 tenths, 6 tenths, 6 hundreds, 6 tenths, 6 thousandths
4 132·415, 3·2651, 96·205, 2·00531, 0·025
5 2 ones/units, 2 tenths, 2 tenths, 2 ones/units, 2 hundreds

C **1** $7 + \frac{3}{10} + \frac{8}{100}$, $6 + \frac{2}{100} + \frac{7}{1000}$;
$12 + \frac{3}{10} + \frac{7}{100} + \frac{5}{1000}$, $7 + \frac{1}{10} + \frac{2}{1000}$
2 0·18 < 0·6, 2·017 > 2·008, 4·17 > 0·417; 0·03 > 0·027, 0·167 < 0·2, 0·076 > 0·009; 1·06 > 1·057, 46·167 < 46·176, 0·6 > 0·36

Page 9 Decimals
A 142·818, 48·527;   17·905, 1551·281;
28·483, 42·358;   48·17, 175·604;   329·175,
126·539
B 50·391, 45·12;   2444·145, 46 153·63;
24 517·343, 450·644;   19·19, 3017·291;
5794·002, 14 386·34
C 1 127·224, 2571·44, 48·204,
1316·42;   10·296, 259·65, 2770·8, 597·104
2 21 605·3, 2387·02, 75 014·4,
29 205·155;   222 987·6, 128 605·08,
89 406·063, 16 930·16;   19 880·496,
156 391·2, 17 025·78, 18 709·8

Page 10 Decimals
A 8·96, 29·93, 19·52, 11·31, 15·01, 30·96;
4·75, 10·36, 10·81, 16·56, 31·02, 14·94
B 37·84, 324·45, 323·26, 66·5, 142·29,
250·95;   225·25, 208·6, 154·96, 386·46,
149·12, 229·71
C 499·59, 1177·29, 875·61, 2005·52,
4852·46;   552·26, 954·32, 1377·51, 2696·54,
10 674·51
D 7·6, 8·9, 5·5, 3·9, 6·2;
0·63, 0·96, 0·47 0·94, 0·56;
13·41, 11·09, 12·58, 16·21, 13·02;
9·471, 4·607, 21·027, 6·473, 2·436

Page 11 Decimals
A 0·126, 0·078, 0·242, 0·528, 0·471;
0·417, 0·303, 0·123, 0·141, 0·214
B 2·35, 3·46, 5·07, 6·34, 1·91;
3·07, 4·61, 6·08, 1·49, 1·42
C 0·842, 0·702, 0·573, 0·641, 0·916;
0·703, 0·435, 0·307, 0·834, 0·537
D 0·399, 0·549, 0·186, 0·282, 0·328;
0·784, 0·204, 0·27, 0·198, 0·273
E 21·165, 41·202, 31·066, 94·134, 87·52
F 73·035, 16·5954, 33·605, 41·195,
185·2617
G 1·143, 116·676, 18·0509;   150·612,
0·1632, 32·3214

Page 12 Decimals
A 0·65, 0·35, 0·72, 0·59, 0·27;
0·34, 0·13, 0·42, 0·07, 0·68
B 5·26, 4·75, 6·03, 9·27, 3·41;
2·98, 7·09, 8·47, 6·25, 5·34
C 3·5, 7·4, 6·5, 8·5, 12·5, 9·8
D 7·35, 5·65, 7·25, 2·75, 9·95, 5·55
E 6·125, 5·375, 9·625, 2·375, 6·875, 8·625
F 5·4, 6·5, 14·5, 9·8, 13·2
G 9·14, 6·25, 7·36, 5·75, 3·94

H 2·125, 5·376, 7·628, 9·875, 7·424
I 4·535, 7·7, 16·35;   7·35, 3·23, 14·8;
13·4, 1·435, 3·446;   12·25, 7·26, 19·74;
6·06, 4·03, 4·51
J 2·21, 0·304, 52·3;   7·25, 0·341, 6·26

Page 13 Decimals
A 12·74, 14·25, 21·21, 34·09, 23·55;
18·07, 32·28, 2·506, 17·35, 13·07
B 4·21, 2·69, 5·29, 27·11, 23·94;
14·22, 6·24, 50·3, 2·07, 1·47
C 8·21, 9·45, 7·32, 5·06 4·27;
6·34, 3·85, 5·25, 4·26, 6·27
D 5·22, 4·73, 6·05, 2·06, 3·42;
1·08, 1·14, 2·04, 1·55, 2·16
E 3·42, 4·73;   69·8, 7·54;   16·27, 14·35;
12·04, 23·09;   72·3, 21·61;   4·52, 6·68

Page 14 Decimals
A 2·1274 **a** 2·127 **b** 2·13 **c** 2·1
3·9681 **a** 3·968 **b** 3·97 **c** 4·0
4·0127 **a** 4·013 **b** 4·01 **c** 4·0
5·3482 **a** 5·348 **b** 5·35 **c** 5·3
6·7296 **a** 6·730 **b** 6·73 **c** 6·7
10·1206 **a** 10·121 **b** 10·12 **c** 10·1
9·0079 **a** 9·008 **b** 9·01 **c** 9·0
6·2396 **a** 6·240 **b** 6·24 **c** 6·2
16·2574 **a** 16·257 **b** 16·26 **c** 16·3
8·3521 **a** 8·352 **b** 8·35 **c** 8·4
B 0·6, 3·2, 1·3, 4·3, 0·2;   1·3, 7·2, 0·8, 1·5,
1·9;   0·3, 3·4, 1·4, 8·2, 4·7
C 0·43, 1·11, 2·10, 0·63, 4·22;   0·91, 0·73,
2·69, 4·73, 3·85;   2·25, 1·43, 2·07, 1·80
D 0·342, 0·453, 0·272, 0·398;   4·722,
2·603, 34·213, 13·032;   0·627, 2·533,
4·301, 1·091
E **a** 1·652 **b** 1·65 **c** 1·7 **a** 0·708 **b** 0·71 **c** 0·7
a 2·042 **b** 2·04 **c** 2·0;   **a** 22·511 **b** 22·51
c 22·5 **a** 2·754 **b** 2·75 **c** 2·8 **a** 1·043
b 1·04 **c** 1·0

Page 15 Decimal and number problems
1 53·93 m **2a** Young **b** Young **c** Ahmed
d Gray, Gray **3** 18 375

Page 16 Decimal and number problems
1a 10 251 **b** £9738.45 **2a** 118 800
b 160 875 **c** 136 125, 356 400 **3** 1848
4 £684.81 **5a** 217 171 **b** 574 492·05
6a 32·96 **b** 174·32 **7** 19 171

Page 17 Equations
A 7, 89;   34, 5;   55, 67;   47, 86;   3, 4
B 29, 100;   6, 28;   31, 12;   64, 24;   25, 7

C 16, 113;　4, 8;　3, 18;　30, 74;　31, 8
D 3, 3;　32, 7;　20, 50;　0, 120;　101, 6

Page 18 Expressions and equations

A $3 \times 7 = 21$, $3 \times 7 \times 9 = 189$, $(2 \times 3) + 3 = 9$, $4 \times 9 = 36$, $2 \times 3 \times 7 + 3 \times 3 \times 9 = 123$; $3 + (5 \times 9) = 48$, $(3 \times 7) + (7 \times 9) = 84$, $(5 \times 7) + (2 \times 9) = 53$, $3 + 7 + (2 \times 9) = 28$, $(6 \times 7) + (3 \times 3) = 51$

B $x = 3$, $x = 5$, $x = 4$, $x = 6$; $x = 10$, $x = 12$, $x = 5$, $x = 5$

C $x - 6$, $4x$, $y + 9$, $\frac{m}{2}$; $r - 6$, $\frac{3h}{10}$, $5 + x$; $2n + r$, $\frac{p}{4} - t$, $v - 15$

D 1 $15 - 3\,r = 3$ 　　 $\frac{x}{3} = 12$ 　　 $\frac{k}{8} = 9$
　　　　 $r = 4$ 　　 $x = 36$ 　　 $k = 72$

　　 $7 + x = 10$ 　 $2x - 15 = 3$ 　 $4x = 24$
　　　　 $x = 3$ 　　 $x = 9$ 　　 $x = 6$

E 5, 5;　4, 7;　6, 24;　8, 30;　35, 30;　5, 2

Page 19 Roman numerals

A 1 25, 9, 23, 27, 19, 24 **2** 40, 81, 77, 44, 84, 43
3 300, 194, 279, 149, 104, 192;　190, 217, 290, 125, 109, 230
4 850, 409, 539, 487, 653;　590, 661, 552, 424, 454
5 1700, 1854, 1321, 1144;　1745, 1602, 1957, 1852
B 1 XXXIX, XXIV, XVI, VII, XXI, XXXIV, XIX, VIII
2 XLVII, LXXIII, LXXXV, LXIV, LVIII, LI, LXIX, XLII
3 CCCXLI, XCVI, CCCXXI, CCLII, CLXXVI, CLIV, CCCXXIX, XCI
4 CDLXXIX, DCCCXXVI, DL, DCCXX, DCXXXV, CDL, DCCCXC, DCCXC
5 MCMXLVII, MDCCCLV, MCDXXXII, MCXXI, MDCCCXL;　MCCCX, MDCCV, MCCLX, MDII, MCCCLXXXV; MDCLXXXVIII, MCDLXVIII, MCMXL, MDCCCXCII
6 XLIII, XCVI, MCXLI, CMII, LVIII, DXX, DCXCIV
7 CCCLIV, MLXXII, XVII, XXXII, MCCXX, LXXXVII, CXLIII
8 MMCCL, MDCII, XXV, MCXI, MMCCXXII, DCCVII, CDXXXI

Page 20 Bases – base 10

A 1 10^4 **a** 10 000 **b** ten thousand; 10^6 **a** 1 000 000 **b** one million; 10^2 **a** 100 **b** one hundred; 10^5 **a** 100 000 **b** one hundred thousand; 10^7 **a** 10 000 000 **b** ten million; 10^8 **a** 100 000 000 **b** one hundred million

2 $10^2 \times 5$ **a** 500 **b** five hundred; $10^3 \times 6$ **a** 6000 **b** six thousand; $10^5 \times 8$ **a** 800 000 **b** eight hundred thousand; $10^4 \times 9$ **a** 90 000 **b** ninety thousand; $10^6 \times 3$ **a** 3 000 000 **b** three million

B 1 36 700, 4 325 000, 41 290, 45 760; 1 114 100, 202 100, 62 300, 35 900; 21 640, 5216, 9009, 82 160
2 2·047, 1141·126, 173, 0·327;　24·53, 1·6, 4·731, 660;　0·432, 32·150, 1·2, 3·941

C 1

	10^6	10^5	10^4	10^3	10^2	10^1	1
3541 × 10			3	5	4	1	0
4561 × 1000	4	5	6	1	0	0	0
4·263 × 1 000 000	4	2	6	3	0	0	0
4 732 140 ÷ 10		4	7	3	2	1	4
1 150 000 ÷ 10 000					1	1	5
6 300 000 ÷ 100 000						6	3

D 1 100 000, 1 000 000, 10 000 000, 10 000
2 100 000, 10 000, 10 000, 100

Page 21 Bases – base 2

A 1 2, 8, 64, 128, 32, 16, 4, 256
2 2^4, 2^7, 2^{10}, 2^5, 2^2, 2^9, 2^3, 2^1 2^8, 2^6
B 1 32, 64, 512, 64, 128
2 4, 4, 8, 8, 2
C 1a 1110, 110111, 101101, 101011, 11011, 110011
b 14, 55, 45, 43, 27, 51
2 27, 42, 7, 9, 17
D 1 111111_2, 101111_2, 11101_2, 100011_2, 10010_2, 111_2
2 101110_2, 11110_2, 111010_2, 110000_2, 111010_2
3 11_2, 10100_2, 1001_2, 11001_2, 1100_2

Page 22 Averages

A 1 5·09 **2** 8·63 **3** 15·86 **4** 3940 **5** 350 527
B 1 346·5 kg **2** £15.75 **3** 672 km **4** 4044 l **5** 147 cm **6a** 116 g **b** 21 g **7a** £26.60 **b** £7.00

Page 23 Money

A 1 23p **a** 27p **b** 77p, £0.47 **a** 3p **b** 53p, 38p **a** 12p **b** 62p, £0.27 **a** 23p **b** 73p, 41p **a** 9p **b** 59p, £0.12 **a** 38p **b** 88p; £0.16 **a** 34p **b** 84p, 48p **a** 2p **b** 52p,

35p **a** 15p **b** 65p, £0.33 **a** 17p **b** 67p,
9p **a** 41p **b** 91p, £0.25 **a** 25p **b** 75p
2 £15.27, £19.02, £17.30, £30.04, £21.60, £19.77
B £12.76, £219.14;   £43.82, £161.13;
£71.93, £374.42;   £85.86, £496.31;
£35.95, £301.01
C £25.28, £52.84;   £66.32, £29.02;
£10.57, £66.87;   £80.87, £4.85;   £55.57,
£13.14

Page 24 Money

A 1 £3.57, £2.91, £5.52, £2.65,
£3.68, £1.53, £2.88
2 4p, 16p, 5p, 6p, 2p, 9p, 7p;
6p, 13p, 11p, 9p, 7p, 8p, 12p;
13p, 11p, 7p, 12p, 9p, 17p, 8p;
4p, 5p, 6p, 7p, 8p, 10p, 9p
B £43.19, £225.54, £10.44;
£110.52, £225.97, £330.88;
£2.55, £26.68, £45.99;
£810.25, £112.32, £313.12
C £6.20, £19.21, £5.09;
£4.23, £11.28, £16.81;
£4.23, £6.07, £1.35;
£2.17, £4.12, £0.22
D £3.24, £5.39

Page 25 Costing

1 48p, £3.80;   48p, 31p;   £1.62, £2.86;
21p, 49p;   12p, £1.69
2

1 kg	500 g	200 g	100 g
£1.00	**50p**	**20p**	**10p**
50p	**25p**	**10p**	5p
40p	**20p**	**8p**	4p
30p	**15p**	**6p**	3p
20p	**10p**	**4p**	2p

35p, 12p;   24p, 6p;   12p, 18p
3 30p, 20p, 30p;   9p, 70p, 49p
4 74p, £1.48, 37p, £3.70, £3.33, £2.59, £7.03,
£4.07
5 50p, £1.00, £4.50, £6.50, £3.00, £2.50,
£7.00, £9.00
6 £4.80, £1.20, £9.60, £19.20, £16.80,
£36.00, £10.80, £13.20, £18.00, £20.40
7 16p, 32p, 48p, 64p, £1.60, £2.40, £2.72,
£3.20, £3.52, £4.00

Page 26 Profit and loss

A 1 £0.75, £1.49, £1.15, £0.99, £1.79, £1.49
2 £1.25, £2.05, £1.75, £1.45, £0.80, £5.85

B 1 £625 **2** £4.35 **3** £12.00 **4** £6.65 **5** £9.75
6a £9.25 **b** £48.75 **7** £62.20 **8** £1068.00

Page 27 Fractions

A 1 $\frac{1}{2}$ **2** $\frac{1}{4}$ **3** $\frac{1}{2}$ **4** $\frac{3}{4}$;
5 $\frac{3}{4}$ **6** $\frac{1}{4}$ **7** $\frac{1}{2}$ **8** $\frac{1}{4}$
B 3 and 6
C 3 and 5

Page 28 Fractions

A 1a $\frac{2}{3}$ **b** $\frac{1}{3}$ **2a** $\frac{1}{4}$ **b** $\frac{3}{4}$ **3a** $\frac{8}{15}$ **b** $\frac{7}{15}$ **4a** $\frac{3}{10}$
b $\frac{7}{10}$ **5a** $\frac{1}{3}$ **b** $\frac{2}{3}$ **6a** $\frac{7}{24}$ **b** $\frac{17}{24}$
B 5, 7, 9, 10, 12, 19, 14, 16;
22, 33, 46, 21, 67, 38, 19, 24;
17, 34, 59, 31, 58, 62, 41, 69;
11, 33, 16, 12, 23, 33, 14, 29

C $\frac{2}{3}$, $\frac{5}{6}$;   2$\frac{4}{5}$, 2$\frac{2}{3}$, 2$\frac{3}{5}$; $\frac{2}{5}$, $\frac{1}{2}$;
6$\frac{2}{3}$, 6$\frac{7}{12}$, 6$\frac{13}{24}$;   1$\frac{1}{3}$, 2; $\frac{5}{8}$, $\frac{7}{16}$, $\frac{3}{8}$

D 3$\frac{3}{4}$, 3$\frac{4}{5}$, 3$\frac{13}{20}$, 3$\frac{7}{10}$;   6$\frac{1}{3}$, 6$\frac{3}{8}$, 6$\frac{1}{4}$, 6$\frac{7}{24}$;   1$\frac{15}{32}$, 1$\frac{9}{16}$,
1$\frac{5}{8}$, 1$\frac{3}{4}$;   4$\frac{3}{10}$, 4$\frac{2}{5}$, 4$\frac{9}{20}$, 4$\frac{7}{15}$;   3$\frac{4}{9}$, 3$\frac{7}{12}$, 3$\frac{2}{3}$, 3$\frac{5}{6}$;
2$\frac{7}{12}$, 2$\frac{11}{18}$, 2$\frac{5}{6}$, 2$\frac{23}{24}$

Page 29 Fractions – addition and subtraction

A 1 8, 6, 10, 20, 16;   10, 20, 24, 18, 21;
30, 12, 12, 35, 24
2 3$\frac{3}{4}$, 2$\frac{7}{10}$, 5$\frac{1}{7}$, 5$\frac{1}{8}$, 2$\frac{1}{2}$, 1$\frac{9}{20}$, 6$\frac{2}{3}$, 2$\frac{1}{11}$
B 1$\frac{9}{40}$, 1$\frac{1}{4}$, 1$\frac{3}{40}$, 1$\frac{7}{40}$;   1$\frac{1}{18}$, 1$\frac{21}{50}$, 1$\frac{11}{20}$, 1$\frac{7}{16}$;
1$\frac{1}{4}$, 1$\frac{5}{8}$, 1$\frac{13}{24}$, 1$\frac{7}{22}$
C 6$\frac{7}{10}$, 6$\frac{3}{16}$, 11$\frac{1}{20}$, 2$\frac{5}{16}$;   2$\frac{7}{10}$, 8$\frac{3}{4}$, 7$\frac{14}{25}$, 7$\frac{17}{18}$;
6$\frac{23}{24}$, 14$\frac{2}{15}$, 11$\frac{11}{20}$, 7$\frac{10}{11}$
D 1$\frac{7}{26}$, 1$\frac{43}{60}$, $\frac{53}{60}$, 1$\frac{16}{45}$;   1$\frac{13}{18}$, 1$\frac{19}{60}$, 1$\frac{55}{72}$, 1$\frac{13}{30}$;
1$\frac{29}{45}$, 1$\frac{35}{72}$, $\frac{37}{42}$, 1$\frac{26}{63}$
E 5$\frac{31}{36}$, 10$\frac{1}{14}$, 15$\frac{59}{80}$, 8$\frac{5}{63}$;   12$\frac{7}{39}$, 14$\frac{53}{60}$, 5$\frac{32}{45}$, 12$\frac{26}{33}$;
12$\frac{23}{60}$, 4$\frac{37}{60}$, 12$\frac{9}{28}$, 8$\frac{1}{60}$
F 20$\frac{13}{40}$, 20$\frac{1}{60}$

Page 30 Fractions – addition and subtraction

A 1 60, 50, 48, 60;   60, 100, 42, 72;
60, 40, 52, 63
2 14, 49, 14, 22;   16, 20, 36, 10
B $\frac{1}{8}$, $\frac{7}{20}$, $\frac{3}{24}$ or $\frac{1}{8}$, $\frac{3}{16}$; $\frac{7}{22}$, $\frac{4}{25}$, $\frac{7}{12}$, $\frac{23}{60}$; $\frac{3}{16}$, $\frac{9}{21}$ or $\frac{3}{7}$,
$\frac{1}{18}$, $\frac{7}{16}$

C $\frac{17}{24}$, $1\frac{17}{20}$, $3\frac{23}{40}$, $5\frac{5}{14}$; $3\frac{19}{20}$, $1\frac{9}{20}$, $3\frac{11}{24}$, $2\frac{13}{24}$; $5\frac{47}{50}$, $1\frac{1}{12}$, $1\frac{7}{12}$, $1\frac{2}{5}$

D $\frac{17}{60}$, $\frac{19}{63}$, $\frac{11}{45}$, $\frac{5}{9}$; $\frac{3}{50}$, $\frac{7}{60}$, $\frac{13}{36}$, $\frac{1}{56}$; $\frac{11}{52}$, $\frac{19}{60}$, $\frac{7}{33}$, $\frac{17}{42}$;

E $\frac{79}{90}$, $2\frac{29}{48}$, $4\frac{59}{60}$, $1\frac{61}{75}$; $\frac{59}{60}$, $1\frac{13}{60}$, $3\frac{4}{27}$, $2\frac{5}{48}$; $4\frac{3}{40}$, $1\frac{27}{50}$, $1\frac{13}{48}$, $1\frac{43}{60}$,

F $4\frac{29}{40}$, $6\frac{7}{12}$, $9\frac{7}{15}$, $8\frac{11}{16}$

Page 31 Fractions – multiplication and division

A 1 $\frac{9}{10}$, $\frac{3}{4}$, $\frac{11}{15}$, $\frac{5}{6}$, $\frac{1}{4}$, $\frac{3}{4}$, $\frac{2}{3}$, $\frac{6}{7}$, $\frac{9}{10}$, $\frac{4}{5}$

2 $\frac{31}{4}$, $\frac{53}{6}$, $\frac{119}{13}$, $\frac{19}{12}$, $\frac{93}{20}$, $\frac{27}{10}$, $\frac{91}{16}$, $\frac{241}{60}$, $\frac{32}{15}$, $\frac{37}{30}$

B $\frac{3}{5}$, $\frac{9}{25}$, $\frac{57}{70}$, $\frac{2}{3}$; $\frac{1}{2}$, $\frac{3}{4}$, $\frac{26}{45}$, $\frac{63}{80}$; $\frac{3}{5}$, $\frac{34}{45}$, $\frac{1}{4}$, $\frac{3}{20}$

C $6\frac{1}{20}$, $5\frac{2}{3}$, 4, $4\frac{11}{20}$; $13\frac{1}{5}$, $9\frac{3}{4}$, $13\frac{1}{3}$, $3\frac{1}{2}$;

D $\frac{9}{10}$, $\frac{4}{25}$, $\frac{2}{3}$, $\frac{3}{8}$; $\frac{27}{200}$, $\frac{2}{105}$, $\frac{3}{8}$, $\frac{28}{33}$

E $\frac{11}{18}$, $8\frac{4}{5}$, 9, 14; 5, $2\frac{3}{4}$, 7, $14\frac{2}{3}$

F $\frac{1}{6}$, $\frac{9}{40}$; $\frac{5}{9}$, $\frac{3}{35}$

G 30, $4\frac{17}{27}$; 7, 57

Page 32 Fractions – multiplication and division

A 15, 21, 12, 5; 7, 20, 8, 3; 25, 12, 22, 12

B $\frac{23}{8}$, $\frac{83}{14}$, $\frac{29}{6}$, $\frac{47}{12}$, $\frac{173}{20}$, $\frac{69}{25}$, $\frac{83}{16}$, $\frac{87}{10}$

C $1\frac{1}{2}$, $\frac{5}{6}$, $\frac{1}{3}$, $\frac{15}{16}$; $3\frac{1}{7}$, $\frac{13}{15}$, $1\frac{7}{8}$, $\frac{3}{10}$; $\frac{14}{15}$, $\frac{1}{4}$, $1\frac{13}{15}$, $2\frac{5}{23}$

D $1\frac{1}{24}$, $\frac{19}{26}$, $4\frac{3}{8}$, $4\frac{4}{9}$; $1\frac{3}{7}$, $\frac{16}{21}$, $1\frac{2}{7}$, $\frac{4}{21}$; $\frac{27}{80}$, $1\frac{13}{20}$, $1\frac{1}{2}$, $1\frac{13}{36}$

E $\frac{9}{10}$, $\frac{5}{6}$, $1\frac{2}{73}$, $1\frac{1}{3}$; 6, $\frac{27}{28}$, $1\frac{5}{7}$, $1\frac{1}{3}$

F $1\frac{1}{15}$, $1\frac{1}{3}$, $\frac{1}{9}$, $2\frac{4}{7}$; $1\frac{5}{28}$, $3\frac{1}{3}$, $5\frac{1}{4}$, $\frac{10}{11}$

G $1\frac{2}{3}$, 6; $3\frac{5}{9}$, 6

H $\frac{2}{15}$, $\frac{3}{4}$; $3\frac{8}{9}$, $\frac{1}{2}$

Page 33 Fractions – mixed examples

A 1 $1\frac{53}{100}$, $1\frac{17}{100}$, $2\frac{7}{20}$, $4\frac{4}{5}$, $14\frac{13}{40}$, $6\frac{3}{100}$; $11\frac{3}{500}$, $15\frac{3}{4}$, $12\frac{31}{50}$, $19\frac{47}{100}$, $\frac{3}{200}$, $3\frac{213}{500}$

2 2·3, 4·35, 6·36, 11·7, 13·8, 6·34, 0·019, 1·85; 5·375, 9·56, 7·65, 6·625, 0·037, 2·034, 1·124, 7·904, 2·875, 3·013, 4·292, 6·776, 6·302, 1·72, 0·55, 2·958

B 1 0·9, 0·95, 0·6, 0·64, 0·125, 0·15, 0·088, 0·78, 0·134, 0·375, 0·45, 0·84

2 4·875, 2·3, 6·4, 8·28, 5·05, 2·408, 4·74, 6·862

C 1 0·636, 0·933, 0·778, 0·813, 0·286, 0·917, 0·833, 0·238, 0·944, 0·667

2 4·385, 2·722, 5·214, 9·576, 6·909, 8·867, 2·688, 1·529

D 1 $\frac{5}{8}$, 0·53, 0·473, $\frac{5}{11}$, $\frac{4}{9}$, 0·432, $\frac{5}{12}$

2 $\frac{3}{8}$, 0·335, 0·326, $\frac{3}{10}$, $\frac{5}{18}$, $\frac{4}{15}$, 0·251

3 $\frac{13}{16}$, $\frac{19}{25}$, 0·755, 0·732, $\frac{8}{11}$, 0·695, $\frac{15}{24}$

Page 34 Fractions – problems

A a £3.05 **b** 8·4 m **c** 385 **d** 5·53 **e** 1·925 km; $\frac{5}{8}$ **a** 200 **b** 2·175 l **c** 1275 **d** 400 g **e** 487·5 mm; $\frac{11}{20}$ **a** 12 122 **b** £9.68 **c** 1·5125 kg **d** 528 m **e** 12·65 t

B a 456 **b** 424 g **c** 990 l **d** 15 **e** 384p **f** 315 m **g** 60 km **h** $8\frac{1}{4}$

C 1 0.05 km or $\frac{1}{20}$ **2** £1.25 **3a** 161 **b** 46 **c** 253

4a £30.00 **b** £20.00 **c** £7.50 **5** $57\frac{37}{60}$ km **6** $\frac{41}{60}$ **7a** 21 **b** $22\frac{1}{2}$ **8** $\frac{53}{112}$

Page 35 Percentages

A 8% = $\frac{2}{25}$, 70% < $\frac{4}{5}$, 35% > $\frac{3}{10}$, 55% = $\frac{11}{20}$; $\frac{3}{4}$ > 65%, $\frac{19}{20}$ > 76%, $\frac{17}{25}$ < 72%, $\frac{3}{5}$ < 62%; $\frac{5}{8}$ < $72\frac{1}{2}$%, $\frac{3}{20}$ > 12%, $\frac{14}{25}$ = 56%, 32% < $\frac{7}{20}$

B 45, 41p, $12\frac{1}{2}$ kg; 12, 18, 97; 13, 22, $7\frac{1}{2}$; 250, 32·5, 12 g; 70p, 4p, 8; 120, 50, 21

C 1a 5 **b** 15 **c** 20 **d** 40 **e** 60 **2a** 50p **b** £1.00 **c** £1.50 **d** £2.00 **3a** 45 **b** 135 **c** 315 **d** 405 **4a** 6 g **b** 18 g **c** 54 g **d** 102 g **e** 138 g **5a** 25p **b** 75p **c** £1.75 **d** £2.25 **e** £2.75 **6a** 125 **b** 375 **c** 625 **d** 875

D 5p, £2.45, $148\frac{1}{2}$ m; 30 g, 21 kg, 39 t; $337\frac{1}{2}$ g, £50, 405p; 288p, 9, 28; 21 cm, 8, £500; £3.00, 110 l, 18

Page 36 Percentages

1 490 **a** 539 **b** 612·5; 750 **a** 825 **b** 937·5; 680 **a** 748 **b** 850; 484 **a** 532·4 **b** 605; 768 **a** 844·8 **b** 960; 924 **a** 1016·4 **b** 1155; **2** 3754 **a** 3003·2 **b** 3566·3; 8902 **a** 7121·6 **b** 8456·9; 7426 **a** 5940·8 **b** 7054·7; 6891 **a** 5512·8 **b** 6546·45; 5249 **a** 4199·2 **b** 4986·55; 7348 **a** 5878·4 **b** 6980·6

3a 10p, 30p, 56p, 13p, 85p **b** £0.85, £2.65, £5.04, £1.12 ,£7.65

4a £21.60 **b** £341.60 **c** £24.30 **5** £22.80

6 £5.00 **7a** £10.47 **b** £147.39 **c** £1469.76

8 12 **9** £1650 **10** 4628

Page 37 Ratio

A 1a 11:20, 9:10, 4:5, 12:15, 21:30, 7:8, 11:12, 19:25, 33:100, 29:50
b 2:5, 7:20, 27:100, 1:8, 3:4, 3:10, 4:25, 7:50
2a $\frac{3}{4}$, $\frac{6}{7}$, $\frac{27}{40}$, $\frac{33}{50}$, $\frac{23}{100}$, $\frac{7}{8}$, **b** 35% 18%, 52%, 72%, 30%, 23%

B 1a 1:2, 2:3, 3:2, 4:3, 3:1, 9:4; 2:3, 4:9, 1:3, 2:1, 3:2, 3:4
2 a 2cm:1.5 cm **b** 2.5 cm:3 cm
c 1 cm:3 cm **d** 2.4 cm:0.6 cm
C 1a 1:5 **b** 1:6 **c** 1:8 **d** 2:3 **e** 3:5 **f** 2:7
g 3:8
2 Check your child's slopes.

Page 38 Ratio and proportion

A 80 m, 42.5 km; 80 m, 21·5 m; 200 km, 74 m; 375 m, 28 km
B 1a 100 g, 500 g **b** 20, 25 **c** 6, 18
d 300 kg, 450 kg **e** 100, 10 **f** 25, 35
2 2:1, 3:4; 7:5, 6:5; 1:2, 4:7; 5:3, 5:3
C 1 140 m **2** 100 m **3** 1:4
D 1 An £20.00, William £5.00 **2a** £250.00
b £375.00 **c** £600.00

Page 39 Capacity

A 1 2·471 l, 3·016 l, 0·743 l; 0·076 l, 0·126 l, 2·354 l
2 2436 ml, 125 ml, 637 ml; 47 ml, 2071 ml, 7 ml
3 $\frac{1}{2}$ l, $\frac{2}{5}$ l, $\frac{9}{20}$ l; $\frac{3}{4}$ l, $\frac{17}{20}$ l, $\frac{7}{20}$ l;
4 $\frac{27}{100}$ l, $3\frac{17}{25}$ l, $1\frac{11}{20}$ l; $6\frac{3}{20}$ l, $2\frac{1}{4}$ l, $10\frac{19}{20}$ l
4 4400 ml, 6250 ml, 700 ml; 350 ml, 1800 ml, 360 ml; 5750 ml, 9956 ml, 452 ml; 2300 ml, 600 ml, 1 600 ml
5 0·38 l > 38 ml, 4·06 l < 4600 ml, $1\frac{3}{5}$ > 1500 ml; $3\frac{3}{10}$ l > 3030 ml, $9\frac{7}{20}$ l > 9·035 ml, 0·035 l < 350 ml
B 1 3·435, 12·926; 5·837, 31·620
2 957, 5·104; 973, 11·777
3 283·278, 15·3; 371·439, 58·88
4 3·216, 5·75; 0·695, 42·937

Page 40 Mass

A 1 2·75 kg, 1·8 kg, 0·7 kg, 0·45 kg, 3·52 kg, 6·027 kg; 3·03 kg, 8·25 kg, 0·5 kg, 0·55 kg, 1·6 kg, 6·013 kg
2 $4\frac{3}{4}$ kg, $\frac{17}{20}$ kg, $\frac{19}{20}$ kg, $3\frac{2}{5}$ kg, $4\frac{7}{10}$ kg; $\frac{37}{40}$ kg, $\frac{1}{20}$ kg, $4\frac{9}{20}$ kg, $9\frac{3}{10}$ kg, $\frac{1}{4}$ kg
3 620 g, 510 g, 0·601 kg **4** 675 kg, 0·6 t,

432 kg **5** 0·75 kg, 2500 g, 2·524 kg, 750 kg, 1 t, 1200 kg
6 3 t 750 kg, 6 t 900 kg, 11 t 750 kg, 6 t 850 kg, 8 t 700 kg, 5 t 800 kg
7 6·027 t, 11·009 t, 4·263 t, 0·347 t, 0·062 t
B 1 10·222 t, 5·129 t; 14·890 kg, 39·16 kg
2 1·60 kg, 0·631 kg; 65·922 t, 1·376 kg
3 50·148 kg, 174 t, 52·056 kg, 351·760 kg
4 3·912 t, 12·125 kg, 22·201 t, 4·4 kg

Page 41 Length

A 1 3·027 km, 21·345 km, 0·432 km, 0·034 km, 0·012 km; 0·053 km, 6·005 km, 2·475 km, 0·026 km, 12·035 km
2 6·26 m, 5·06 m, 21·027 m, 0·045 m, 0·096 m; 12·175 m, 16·07 m, 2·141 m, 21·006 m, 0·007 m **3** 6·25 km, 11·5 km, 17·75 km, 3·7 km, 2·8 km, 0·55 km; 8·6 m, 12·3 m, 2·35 m, 12·75 m, 11·45 m, 14·9 m
4 $12\frac{1}{2}$ km, $26\frac{3}{4}$ km, $32\frac{7}{10}$ km, $45\frac{1}{20}$ km, $16\frac{3}{200}$ km; $9\frac{3}{5}$ m, $13\frac{9}{20}$ m, $\frac{1}{4}$ m, $18\frac{11}{20}$ m, $\frac{19}{20}$ m
5 14 km 750 m, 6 km 234 m, 8 km 632 m, 12 km 275 m, 12 km 700 m; 6 km 450 m, 10 km 35 m, 11 km 800 m, 8 km 900 m, 9 km 600 m
B 1 7·965 km, 7·032 km, 33 cm 7 mm, 11·4 cm
2 9·5 cm, 98·568 km; 12·738 m, 114·805 km;
3 15 m, 24·750 cm; 1404 cm 7 mm, 657 m 81 mm; 100·275 km, 1·875 m
4 39·3 cm, 46·5 m; 0·396 km, 87·3 cm; 4·56 m, 11·251 m

Page 42 Measures – problems

1 90 **2** 88·13 cm **3** 17·4 l **4** 2165 l
5 8117·2 km **6** 904 kg **7** 2400 kg **8** 9600
9 61·25 m **10** 1620 l **11** 17 122 km
12 43 t 188 kg

Page 43 Area and perimeter

A 1 area 525 mm², perimeter 100 mm
2 area 884 mm², perimeter 120 mm
3 area 611 mm², perimeter 120 mm
B a 54·8 cm **b** 102·8 mm **c** 43·2 m;
d 42 cm **e** 2090 mm **f** 53·8 m
C length **19 cm, 17 cm, 26 cm**;
width **11 cm, 15 cm**;
perimeter **68 cm, 54 cm, 70 cm, 118 cm, 98 cm**; area **285 cm², 228 cm², 375 cm², 850 cm²**
D 1a 58 mm **b** 125 mm² **2a** 96 mm

b 312 mm^2 **3a** 110 mm **b** 460 mm^2
E 127·5 cm^2, 141 cm^2;   220 cm^2, 33·25 m^2;
1012·5 mm^2, 75·25 cm^2
F base **15 cm**;   **10 cm**;
height **10 cm**, **12 cm**, **8 cm**;
area **120 cm^2**, **108 cm^2**

Page 44 Area and perimeter
1 1387·5 cm^2 **2** 9 m^2 **3** 826·5 m^2
4 10·56 m^2
5 17·55 m^2 **6** 1095 m^2 **7** 8·225 m^2
8 405 mm^2

Page 45 Area
1 880 cm^2 **2** 1550 m^2 **3** 8·4 cm^2 **4** 2706 m^2
5 1976 mm^2 **6** 1562·5 m^2 **7** 2100 km^2
8 2175 cm^2

Page 46 Circles – circumference
A **1** Check your child's circles.
2 Check your child's circles.
B **1** 5·6 m, 4·8 cm, 6·35 m, 7·6 m, 8·45 cm,
6·75 cm
2 $15\frac{1}{2}$ cm, 31 cm, $4\frac{1}{2}$ m, $13\frac{2}{5}$ m, $9\frac{1}{5}$ cm, 25 m
C **1** $17\frac{2}{7}$ cm, $42\frac{3}{7}$ cm, $28\frac{2}{7}$ cm, $56\frac{4}{7}$ cm,
$78\frac{4}{7}$ cm, $45\frac{4}{7}$ cm
2 22 cm, $20\frac{3}{7}$ cm, $47\frac{1}{7}$ cm, $59\frac{5}{7}$ cm,
$24\frac{5}{14}$ cm, $45\frac{4}{7}$ cm
D **1** 16·956 m, 21·352 cm, 43·332 m,
116·18 cm, 104·876 cm
2 93·572 m, 148·208 cm, 111·784 cm,
58·09 m, 2·355 m

Page 47 Circles – area and circumference
A **1** 154 cm^2, 616 cm^2, 2464 cm^2, 3850 cm^2,
$86\frac{5}{8}$ cm^2, $194\frac{29}{32}$ cm^2
2 $9\frac{5}{8}$ cm^2, $86\frac{5}{8}$ cm^2, $346\frac{1}{2}$ cm^2, $471\frac{5}{8}$ cm^2,
$779\frac{5}{8}$ cm^2, $962\frac{1}{2}$ cm^2
B **1** 28·26 cm^2, 254·34 cm^2, 113·04 cm^2,
379·94 cm^2, 50·24 cm^2, 153·86 cm^2
2 615·44 cm^2, 452·16 cm^2, 1017·36 cm^2;
379·94 cm^2, 1808·64 cm^2, 78·5 cm^2
C **1** Diameter (in cm) **10·2, 9·0, 10·0, 2·8,
16·0**;
Radius (in cm) **6·4, 4·5, 5·0, 0·8, 8·0**;
Circumference (in cm) **40·192, 32·028,
31·4, 8·792, 5·024**;
Area (in cm^2) **128·6144, 81·6714, 63·585,
6·1544, 2·0096, 200·96**
2a area 1595·3125 cm^2, 172·25 cm
b area 557 cm^2, 119·4 cm

Page 48 Volume of cubes and cuboids
A **1** 125 cm^3
2 64 cm^3, 216 cm^3, 1331 cm^3, 729 cm^3,
1000 cm^3, $91\frac{1}{8}$ cm^3;
$274\frac{5}{8}$ cm^3, $1157\frac{5}{8}$ cm^3, $1520\frac{7}{8}$ cm^3,
$1953\frac{1}{8}$ cm^3, $614\frac{1}{8}$ cm^3
B **1** 240 cm^3
2 126 cm^3, 126 cm^3, 217 cm^3, 90 cm^3,
$173\frac{1}{4}$ cm^3, $312\frac{1}{2}$ cm^3, 420 cm^3
3a $l = \frac{v}{b \times h}$ **b** $b = \frac{v}{l \times h}$ **c** $h = \frac{v}{l \times b}$

Page 49 Volume – triangular prisms and cylinders
A **1** x $13\frac{1}{2}$ cm^3, y 64 cm^3 **2a** 150 cm^2,
4500 cm^3 **3** 8640 cm^3
4 225, $202\frac{1}{2}$, 700, $1068\frac{3}{4}$, 3450, 2000, 3360,
11 250, 4200, 4125
B **1a** $314\frac{2}{7}$ cm^2 **b** $9428\frac{4}{7}$ cm^3
2 $4525\frac{5}{7}$ cm^3
3 502·4, 3077·2, 141·3, 3052·08, 100·48,
3391·2

Page 50 Volume – assorted shapes
1 19 250 cm^3 **2** 5400 cm^2 **3** 12 250 cm^3
4 60 000 cm^3 **5** 7680 cm^3 **6** 5775 cm^3
7 6600 m^3 **8** 22 850 cm^3

Page 51 Scale drawing
A **1a** Manchester 2·9 cm, Liverpool 3 cm,
Southampton 2·5 cm;
Edinburgh 6 cm, Aberdeen 7·8 cm
b Penzance 450 km, Glasgow 620 km,
Cardiff 250 km;
York 310 km, Leeds 300 km
B **a** 2 m **b** 1·4 m **c** 40 cm **d** 30 cm
e 58 cm **f** 50 cm **g** 26 cm **h** 34 cm
j 46 cm **k** 90 cm **m** 60 cm **n** 66 cm
C 1 cm:20 km, 1 cm:2 km, 1 cm:50 km,
1 cm:40 km, 1 cm:25 km

Page 52 Scale drawing
A 15 m – 3 cm, 20 m – 4 cm, 5 m – 1 cm,
10 m – 2 cm;   25 m – 5 cm, 10 m – 2 cm,
15 m – 3 cm
B **a** 110 km **b** 200 km **c** 320 km **d** 480 km
e 415 km **f** 510 km **g** 420 km **h** 310 km
i 175 km **j** 90 km
C **1** A 230 km B 330 km C 220 km
D 350 km E 270 km F 230 km G 300 km
H 420 km I 550 km J 160 km
2a A 115 km B 165 km C 110 km D 175 km
E 135 km F 115 km G 150 km H 210 km
I 275 km J 80 km **b** A 46 km B 66 km

C 44 km D 70 km E 54 km F 46 km
G 60 km H 84 km I 110 km J 32 km
c A 92 km B 132 km C 88 km D 140 km
E 108 km F 92 km G 120 km H 168 km
I 220 km J 64 km **d** A 460 km B 660 km
C 440 km D 700 km E 540 km F 460 km
G 600 km H 840 km I 1100 km J 320 km

Page 53 Angles

A 1 Check the angles your child has drawn.
2a 120°, 60° **b** 140°, 40° **c** 105°, 75°
d 30°, 150° **e** 110°, 70° **f** 160°, 20°
g 80°, 100° **h** 15°, 165° **3a** 145° **b** 115°
c 50° **d** 35° **e** 155° **f** 125° **4a** 163° **b** 108°
c 80° **d** 95°
B 1a p 40°, q 140°, r 40°, s 140°
b p 45°, q 135°, r 45°, s 135°
c p 30° q 150°, r 30°, s 150°
2a x 160°, y 20°, z 160° **b** a 148°, b 32°
c 148° **c** r 124°, s 56°, t 124°

Page 54 Angles

A 1a X 90°, Y 30°, Z 60° **b** X 85°, Y 50°, Z 45°
c X 90°, Y 35°, Z 55°
2a Check the angles your child has drawn.
b 110°, 55°, 75°, 45°, 85°; 30°, 35°, 25°, 65°,
50°
3a 74° **b** 17° **c** 35° **d** 46°
B 1a P 80°, Q 115°, R 60°, S 105° **b** P 140°,
Q 95°, R 75°, S 50° **c** P 110°, Q 85°, R 65°,
S 100°
2 80°, 125°, 135°; 50°, 105°, 85°
3 100°, 20°, 30°

C

depart	09:00	09:25	09:50	10:15	10:40	11:05	11:30	11:55
arrive	09:20	09:45	10:10	10:35	11:00	11:25	11:50	12:15

Page 57 Speed

A 1a 300 km **b** 390 km **c** 495 km **d** 165 km;
e 258 km **f** 132 km **g** 275 km **h** 110 km
2a 4 h 10 min **b** 5 h 20 min **c** 4 h 22 min
d 7 h 35 min **e** 10 h 10 min
B 1a 240 h **b** 96 h **c** 48 h **d** 160 h;
e 60 h **f** 32 h **g** 80 h **h** 40 h
2a 96 km/h **b** 64 km/h **c** 80 km/h
d 120 km/h **e** 60 km/h
C 1a 27·5 km/h **b** 51 km/h **c** 1050 km/h
d 183·75 km/h **e** 420 km/h **f** 61·5 km/h
2a 10 m/s **b** 40 m/s **c** 10 m/s **d** 10 m/s
e 20 m/s **f** 6 m/s
3a 360 km/h **b** 828 km/h **c** 691·2 km/h
d 540 km/h **e** 756 km/h **f** 288 km/h

Page 55 Bearings

A Check your child's bearings.
B 070°, 500 km   290°, 600 km
270°, 400 km   127°, 660 km
110°, 720 km   050°, 660 km
280°, 740 km   235°, 560 km
C 145°, 420°, 250°, 330 km
D

name	bearing		distance	
	a	b	a	b
Peter	080°	55°	1·4 km	0·4 km
Freya	070°	350°	1·1 km	0·4 km
John	055°	000°	1·3 km	0·7 km
Olivia	040°	320°	0·8 km	0·8 km
Alana	010°	295°	0·5 km	1·1 km
Jack	320°	290°	0·7 km	1·6 km

Page 56 Timetables

A 1 09:00, 09:40, 10:20, 11:00, 11:40, 12:20,
13:00, 13:40, 14:20, 15:00
2 10:15, 10:55, 11:35, 12:15, 12:55, 13:35,
14:15, 14:55, 15:35, 16:15
3 10:30, 11:10, 11:50, 12:30, 13:10, 13:50,
14:30, 15:10, 15:50, 16:30
4 11:45, 12:25, 13:05, 13:45, 14:25, 15:05,
15:45, 16:25, 17:05, 17:45
B 1 depart 11:15, 12:15, 13:15, 14:15, 15:15,
16:15
arrive 10:55, 11:55, 12:55, 13:55, 14:55,
15:55, 16:55
2 8 **3** 16:15 **4** 14:15 **5** 576 **6** £192

4a 50 m/s **b** 15 m/s **c** 35 m/s;
d 45 m/s **e** 7·5 m/s **f** 70 m/s

Page 58 Speed

A 1a average speed 1000 km/h **b** average
speed 960 km/h **c** distance 4000 km
d average speed 840 km/h
e time 3 h 50 min
f average speed 800 km/h
2 b and e **3** c, a, b and e, d, f
B 1 Mach 2 = 2400 km/h,
Mach 3 = 3600 km/h,
Mach 4 = 4800 km/h, Mach 5 = 6000 km/h,
Mach 6 = 7200 km/h;
Mach 2·5 = 3000 km/h,

Mach 4·75 = 5700 km/h,
Mach 3·2 = 3840 km/h,
Mach 4·1 = 4920 km/h,
Mach 6·3 = 7560 km/h
2 7440 km/h = Mach 6·2,
2880 km/h = Mach 2·4,
1980 km/h = Mach 1·65,
5700 km/h = Mach 4·75;
6300 km/h = Mach 5·25,
3960 km/h = Mach 3·3,
2850 km/h = Mach 2·375
1320 km/h = Mach 1·1

Page 59 Graphs
A **1a** Friday **b** Sunday
2 53 **3** 48 **4** Sunday **5a** Saturday **b** Tuesday
6 Saturday **7** 4 **8** 3
B Check your child's graph.
C kilometres 5 mi = **8 km**, 15 mi = **24 km**,
30 mi = **48 km**, 40 mi = **64 km**;
$17\frac{1}{2}$ mi = **28 km**, $27\frac{1}{2}$ mi = **44 km**,
$42\frac{1}{2}$ mi = **68 km**, $22\frac{1}{2}$ mi = **36 km**
miles 24 km = **15 mi**, 48 km = **30 mi**,
32 km = **20 mi**, 56 km = **35 mi**;
20 km = **$12\frac{1}{2}$ mi**, 52 km = **$32\frac{1}{2}$ mi**,
12 km = **$7\frac{1}{2}$ mi**, 60 km = **$37\frac{1}{2}$ mi**,
D Check your child's conversion graphs.

Page 60 Comparison graphs
A **1** Sunday 1500, Monday 2800, Tuesday
5000, Wednesday 3700;   Thursday 3500,
Friday 4500, Saturday 800

2 Sunday 900, Monday 2000, Tuesday
4400, Wednesday 4000, Thursday 2500,
Friday 3500, Saturday 3500
3 21 800 **4** 20 800 **5** 1000 **6** Tuesday,
Wednesday, Friday
B Check your child's graph.
C **1a** 20 km/h **b** 30 km/h **2a** 5 km, $7\frac{1}{2}$ km
b 10 km, 15 km **c** 25 km, $37\frac{1}{2}$ km **d** 30 km,
45 km **e** 35 km, $52\frac{1}{2}$ km **3a** 2 km **b** $3\frac{1}{2}$ km
c $6\frac{1}{2}$ km **d** $11\frac{1}{2}$ km **e** $13\frac{1}{2}$ km
f $19\frac{1}{2}$ km **4** 30 min
D Check your child's graph.

Page 61 Co-ordinates
A **1** a (1, 2) j (4, 4) g (5, 7) c (7, 6) d (8, 2)
k (9, 3) b (11, 5) h (13, 2) f (15, 5) l (15, 1)
i (18, 7) e (19, 3)
2 Check your child's plotted co-ordinates.
B **1** Check your child's co-ordinates.
2a T **b** H **c** I **d** N **e** K

Page 62 Co-ordinates
A **1** 2, 4, 6, 8, 10, 12, 14, 16, 18, 20, 22, 24
2 3, 15, 21, 13;   19, 7, 11, 23;
$7\frac{1}{2}$, $10\frac{1}{2}$, $11\frac{1}{2}$, $4\frac{1}{2}$; $2\frac{1}{2}$, $6\frac{1}{2}$, $8\frac{1}{2}$, $9\frac{1}{2}$
B **1** x 10, 25 y 20, 60, 80, 120
2 48, 112, 72, 88, 64, 104
3 $17\frac{1}{2}$, 9, 27, $21\frac{1}{2}$, $12\frac{1}{2}$, 7
4 36, 84, 116, 64;   52, 112, 76, 44;
27, 23, 19, 29, 14, 26
C Check your child's graph.